DOC029694

MANHATTAN DISTRICT HISTORY

BOOK I - GENERAL

VOLUME 1 - GENERAL *App. A-D*

~~SECRET~~ Copy #3 *A*

REDACTED COPY

2013 00006684

REVIEWED AND NOT DECLASSIFIED
BY U. S. DEPARTMENT OF ENERGY
OFFICE OF CLASSIFICATION
JOHN A. HARTSOCK
REVIEWED BY

9/4/79
DATE

25441

H3753

MDH - BkI - VoL I, App A-D

THIS DOCUMENT CONSISTS OF ....165... PAGE(S)
NO. ......3..... OF ....4.... COPIES, SERIES ...A......

MANHATTAN DISTRICT HISTORY

BOOK I - GENERAL

VOLUME 1 - GENERAL

FOREWORD

This Volume 1 of Book I of the Manhattan District History introduces the reader to the books and volumes that follow. It describes the important published fore-runners of the history, the purpose and scope of the History, and the handicaps under which it was written; and then it describes how it is formed and (very briefly) what it contains. A brief paragraph tells what is to be found in each of the other volumes of Book I, which, because of their subjects, must be more or less individually independent; then, with similar brevity, a few words of description of each of the other books of the History (Book II through Book VIII) are given - just to give the reader an idea of what main subject can be found treated in each one.

Following the introductory section, this volume describes certain general subjects which apply to the whole of the History and could not be suitably included in any other volume or book: "The Mission of the Manhattan District"; "Authorizations and Lines of Authority", and "Total Costs".

From its very nature, this volume must differ in some respects from other volumes of the History: it contains no summary; its appendix references (Appendix A) are confined to the section dealing with authorizations only. (The references and the appendix data which apply to the statements in the introduction and elsewhere in this volume can be found in the later volumes and later books which are described.)

Other appendices (B, C and D) contain: a combined Table of Contents of all the eight books of the History; Combined Indices of the names of persons and organizations mentioned; and a Map showing the geographical locations of the Manhattan District installations and offices. The locations shown on the map in Appendix D include only those at which the District maintained offices, and personnel assigned to duty officially for considerable periods. The locations of operations in which the Manhattan District had a direct interest were located in nearly every state in the Union and in many foreign areas as well - in such widely scattered places as: the Belgian Congo; Bikini Atoll in the Marshall Islands; Great Britain; Canada; Italy; France; Germany; Tinian in the Marianas; and Hiroshima and Nagasaki in Japan. Appendix E contains a general index of the History.

As explained in the introductory section of this volume, the official termination date of the History was 31 December 1946, the day on which the Manhattan District turned over its control of atomic energy matters to the Atomic Energy Commission. Wherever a writer of the History has used the phrase "termination of the Manhattan District", or words to that effect, it is intended to refer to the termination of the District's control (on 31 December 1946) and not to the end of the existence of the District. Actually, although this was not generally understood, the Manhattan District continued in existence for some months longer (during the process of liquidation), until it was officially abolished on 15 August 1947.

December 1948

TABLE OF CONTENTS

SECTION 3 - AUTHORIZATIONS AND LINES OF AUTHORITY

MANHATTAN DISTRICT HISTORY

Book I, General - Volume 1, GENERAL

SECTION 1 - INTRODUCTION TO THE HISTORY

1-1. Published Forerunners of the History.

a.   The President's Announcement. - The first words to reach
the general public - in this or any other country - of the successful
development of the atomic bomb, were contained in the announcement of
President Harry S. Truman on 6 August 1945.

The statement issued by the President on that day announced
that "an American airplane had dropped one bomb" on the city of Hiroshima
in Japan and "destroyed its usefulness to the enemy". After a few
descriptive words on "the battle of the laboratories", which had produced
the bomb, and about the pooling of the interests of the United States
and Great Britain, the President went on to say: "The greatest marvel
is not the size of the enterprise, its secrecy, nor its cost, but the
achievement of scientific brains in putting together infinitely complex
pieces of knowledge by many men in different fields of science into a
workable plan. And hardly less marvelous has been the capacity of
industry to design, and of labor to operate, the machines and methods
to do things never done before, so that the brain child of many minds
came forth in physical shape and performed as it was supposed to do .
Both science and industry worked under the direction of the United
States Army, which achieved a unique success in managing so diverse a
problem in the advancement of knowledge in an amazingly short time. It
is doubtful if such another combination could be got together in the

world. What has been done is the greatest achievement of organized
science in history. It was done under high pressure and without failure."

      b. <u>War Department Announcements</u> - Immediately after the
President had issued his announcement, other statements were released
by the War Department, the first by Secretary of War Henry L. Stimson,
in which the general history of the atomic bomb project through the war
years was briefly outlined. These statements were confined to the bare
bones of history, mentioning the parts played in the earlier scenes by
the pioneering nuclear scientists of many countries and, in particular,
by those of the United States, Great Britain and Canada; then, the parts
played by President Franklin D. Roosevelt; by the Office of Scientific
Research and Development; and by the "General Policy Group" and the
"Military Policy Committee", headed by Dr. Vannevar Bush and Dr. James B.
Conant. These statements also described how decision was made and approved
by the President, in June 1942, to expand the work greatly and to transfer
the program to the War Department; how the enterprise then received the
name "Manhattan Project"; how the organization carrying on the work was
named the "Manhattan District"; how Major General L. R. Groves was
appointed by the Secretary of War to take complete executive charge of
the program and was made directly responsible to the Secretary of War
and the Chief of Staff; and how a total of about two billion dollars had
been spent.

      The statement issued by the Secretary of War also disclosed,
by brief descriptions, the principal locations at which the work of the
Manhattan District had been carried on: the Clinton Engineer Works and
the "city" of Oak Ridge, "on a Government reservation of some 59,000

acres, near Knoxville, Tennessee"; the Hanford Engineer Works and the town of Richland, "on a Government reservation of 430,000 acres" near Pasco, Washington; and the "special laboratory dealing with the many technical problems involved in putting the components together into an effective bomb...in an isolated area in the vicinity of Santa Fe, New Mexico" (Los Alamos). The statement also referred to other smaller plants in the United States and Canada, which had produced needed materials, and it listed a few of the many laboratories - at Columbia University, University of Chicago, University of California, Iowa State College, - which contributed materially to the research and development of special equipment, materials, and processes for the project. It mentioned likewise, by name, just a few of the many industrial organizations which had borne such a vital part in the success of the project; E. I. du Pont de Nemours & Company, M. W. Kellogg Co., J. A. Jones, Co., Union Carbide & Carbon Co., Stone and Webster Engineering Corp., Tennessee-Eastman Co.; and equipment manufacturers: Allis-Chalmers, Chrysler, General Electric, Westinghouse. There were many others which might have been mentioned also.

     c. "The Smyth Report" - Then, a few days later, on 12 August 1945, after the second combat atomic bomb had been dropped on Nagasaki (on 9 August 1945) and two days before Japan offered to surrender, the official report entitled "Atomic Energy for Military Purposes", by Professor H. D. Smyth of Princeton University, was issued. This report, which was carefully scrutinized and censored, before publication, by special appointees of General Groves, to insure that it would not disclose any secret information which might imperil the security of the

United States or its partners in the enterprise, goes into considerable
detail in describing the basic scientific theories and facts, appli-
cable to the project, which were generally known to the world of science
before the project was undertaken. It also describes, in general terms,
the problems which had to be, and were, solved; the general progress
which was made at successive stages; the various methods and processes
which were considered and undertaken for the production of fissionable
material; and the administrative organizations which directed and controlled the work. (The preparation and publication of the Smyth Report
are described in Chapter 13 of Volume 4 of this book, and the Report
itself may be found among the press releases in Chapter 8 of the same
volume.)

Dr. Smyth starts, for his text, with Dr. Albert Einstein's
formula for the equivalence of mass and energy:

$$E = MC^2$$

(where "E" is energy, "M" is mass, and "C" is the velocity of light.
If mass is expressed in grams, and velocity of light in centimeters per
second, energy will be expressed in ergs.) He describes, step by step,
the discoveries of important phenomena of radioactivity and their
significance, leading up to the discovery of uranium fission and its
implications, until, in June 1940, although a chain reaction had not
yet been obtained, "its possibility - at least in principle - was clear

and several paths that might lead to it had been suggested". * (Smyth 1.56.)

It was at about this time, as described by Dr. Smyth, that a censorship committee was set up, in the National Research Council, to control publication of papers on uranium fission; and then, after the Manhattan District was placed in charge, in the summer of 1942, responsibility for promulgation and enforcement of all security regulations relating to the entire project was delegated to the Commanding General of that organization. Very little information - and none of it of serious import - leaked out, so that the project well earned the title of "the best-kept secret of the War".

d. Later Publications - All developments, from the summer of 1940 until August 1945, were carried out behind the curtain of secrecy imposed by war time conditions. As we have seen, in August 1945 the curtain was lifted slightly, first by the President and then by statements issued by the War Department, including Dr. Smyth's report. This was after successful combat use of the atomic bomb, when it was felt to be militarily useful to apprise the Japanese as fully as possible of the enormity of the weapon which was being used against

*(Footnote: It was in April 1940, that the book by Dr. Harvey E. White, of the University of California, "Classical and Modern Physics", was first published, with a preface dated January 1940, and a chapter on the Atomic Nucleus containing the following statements (pp. 615,616): "...uranium may turn out to be an available source of untold energy"; and: "should uranium atoms of weight 235 be responsible for the observations just described, it seems reasonable to suspect that a small piece of uranium metal, composed entirely of these atoms alone, should act like a bomb and explode with far greater violence than any known explosive.")

them. From time to time during the period following August 1945 the
curtain has been lifted further, for the most part by statements issued
by the War Department and articles, papers, etc., approved by that
Department, but the vital secrets of just what the Manhattan District
did, and how, have remained undisclosed to the general public.

1-2. Purpose and General Scope of the History.

a. What, How, When, Where - The Manhattan District History,
most of which is, and may remain, a classified secret document, is intended
to describe, in simple terms, easily understood by the average reader,
just what the Manhattan District did, and how, when, and where. It
describes, briefly, the various parts of the organization and who did
what; it shows, in round figures, how much the various parts of the work
cost, how they were authorized; how contractors were selected, and how
contracts were made, administered, and carried out. Every effort has
been made to insure that the document as a whole is historically correct,
as to facts, theories and opinions. Insofar as the historical evaluation
of the significance of all these facts, theories and opinions is concerned, sufficient time has not elapsed between the occurrence and the
record. Such analysis must be left for future historians to perform,
but the present purpose has been to include, as far as possible, all
significant data, showing all sides of every pertinent question. Not
only huge successes but also minor failures of the Manhattan District
will be found recorded in these pages. Uncertainties, difficulties,
occasional fumblings have been included and, in some instances, where
differences of opinion on any important subject have existed, the attempt
has been made to give the reader two or more different views.

b. <u>Evaluation of Credit</u> - Among the matters which cannot be evaluated properly and fairly at this time is the degree of credit for success - either in large or in small areas of the activities of the Manhattan District - to which any individual or any group of individuals may be entitled. Even if the perspective of the recent past had already become sufficiently defined to permit meritorious work to be evaluated fairly, the mechanics of the writing of this History would render it difficult, if not impossible, to attain uniformly correct results. The History is the work of many writers and no two of them could have the same understanding of the project as a whole; no two of them could attain the same comparative ratings in evaluating the deserts of individuals engaged on widely separated parts of the project; one writer, from a limited viewpoint, might overpraise a minor accomplishment, while another writer might underestimate the work of a major accomplishment. Accordingly, every effort has been made to prevent the writers from including in the History any evaluation of the accomplishments of individuals or groups of individuals. Insofar as possible, the names of those concerned in every part of the project have been recorded, but words of praise have been eliminated. The unadorned factual recital of accomplishments should thus permit each reader to exercise his own judgment, if he wishes, in determining the degree of credit to which he thinks various individuals are entitled; and those who read the record in future years, after time has arranged all events in proper perspective, may be able to determine more fairly and accurately the relative positions in history of those who shared in this enterprise.

c. _Introductory Material_ - In telling the story of the Manhattan District, the History includes, as background information or as introductory material, much of the history of events which transpired before the District was organized. This applies most particularly to the vital history-making activities of the Office of Scientific Research and Development, prior to the beginnings of the District in 1942 and until that Office turned over the last of its contracts in the field of atomic energy to the District in 1943. It applies also to much of the early scientific history which is so well covered by Dr. Smyth in his report. Although the Smyth Report has been incorporated as a part of the Manhattan District History, nearly everything in it is repeated, in greater detail, in one volume or another of this History. Indeed, because the same or similar data must serve as introductory material to more than one phase of the subject (as, for example, various different processes of isotope separation) some repetition within the History itself has been unavoidable. Especially because different subjects have been treated by different writers, it has been felt that the advantages of coherence and completeness in treating each subject separately will outweigh the disadvantage of occasional repetitions.

d. _Period Covered_ - The period covered by the History, then, may be said to start _prior_ to the date of the official activation of the Manhattan District on 16 August 1942. This period extends through the successful use of the atomic bomb on Japan, through the termination of hostilities, to include all the activities of the Manhattan District which followed, until the entire project was officially turned over to the United States Atomic Energy Commission, in accordance with the terms of the Atomic Energy Act of 1946, on 31 December 1946.

1-3. Handicaps of the History.

a. Early Attitude Toward Records — At the beginning of the project, when a major purpose of the work of the Manhattan District was to beat the Germans in reaching the goal of an effective atomic bomb, while insuring in every conceivable way that the enemy should not find out what was going on in this country, the attitude of all concerned was one of general opposition to records. Records were dangerous. They must be dispensed with or destroyed whenever possible. The logical consequence was that original records of actions, thoughts, purposes, discussions or decisions were in some respects non-existent or meager in the extreme. This has been a difficult handicap to the writers of the History. When the need for writing the history was realized, at about the end of 1944, it became necessary to dig up from here, there and everywhere such incomplete records as were available and to supplement them, oftentimes, by the recollections of individuals concerned.

b. Compartmentalization — Another serious handicap arose from the policy of "compartmentalization" which governed the security regulations under which the Manhattan District was operated. Although the successful protection of the secrecy of the project was in no small measure due to this policy, it cannot be denied that compartmentalization made the history-writing task more difficult. With those concerned with the project permitted to know only so much of what was going on as they required for their own specific parts of the work, it became necessary to use many writers and for these writers to consult many people, to piece the various parts of the story together. A further result of this situation was difficulty in determining the relative amount of weight to be attached to different pieces of the story. If some parts of the History appear to be unbalanced, with undue weight or an undue amount of space given to one subject and a corresponding inadequacy in the treatment of another, compartmentalization has been in some measure responsible.

c. <u>Early Outline of the History</u> - Compartmentalization also handicapped the early planning of the History. Those who first attempted to make a workable outline, to form a framework which could be uphol-stered, bit by bit, by the stories of the various activities of the District, were themselves ignorant of many of the phases of the work which would be described. It was therefore necessary to revise the outline itself from time to time, as the writing progressed, and to move some parts of the history from one place to another, as from one volume to another, or from one book to another. This did not result in an ideal ultimate arrangement and such faults as there are in the general layout and arrangement of the books and volumes, the chapters, sections and paragraphs, have been caused in some measure by this condition. For one notable example, two volumes of Book I, Nos. 2 and 3, have been omitted altogether, because these volumes were originally reserved for material which was ultimately placed elsewhere.

d. <u>Words vs. Deeds</u> - The important beginnings of the writing of the History were undertaken during the first half of 1945, while the project was at the peak of production activity, when all were straining every nerve to attain their goal - to <u>do</u>, to accomplish results.

Under these conditions, it is not surprising that many of those whose help in writing the History was invaluable could not spare the time to make such help as complete as it might have been. Deeds necessarily took precedence over words.

e. Turnover of Key Personnel - When hostilities ceased, there were wide gaps in the History still to be filled and at that time numerous changes in personnel began, thus adding a further handicap. In the absence of key personnel who may have been thoroughly familiar with actions of the past, the difficulty of obtaining information about these past actions was measurably increased.

f. Extension to Termination Date - When the writing of the History was first undertaken, full-scale hostilities were proceeding against our two major enemies, and the activities of the Manhattan District were still continuing into an indefinite future. There were no means of determining the termination of the period to be covered by the History. Few, if any, of the volumes were commenced on coincident dates, and each writer was requested to prepare his draft to some convenient date in the near past, as of his particular date of writing; the termination dates of the original drafts of different volumes ranged from April 1945 through December 1945, with some bearing the date of the end of the fiscal year, 30 June 1945, and some bearing a date near the end of hostilities, July or August 1945. The expectation was that periodic supplements or extensions of the volumes would be written in the future, after a final draft of the original had been completed. When, in December 1946, the date of termination of the period of the History, 31 December 1946, had been determined, it became necessary to arrange for all volumes, both those which were finished and those which were unfinished, to be continued to that date. The policy adopted, in general, has provided for extending supplements for those volumes which had been finished, or nearly finished, when this date was fixed; the other volumes were revised so as to cover the period of extension in the main text, all the way through. These circumstances added somewhat to the complications to which the writing was subjected and increased materially the time required to complete the History. They also explain one of the features in which the finished volumes depart from uniformity .

g. No Apology - The enumeration of the above handicaps is not intended as an apology. It is believed that the History as a whole can and should be judged on its merits as it is, without comment of praise or criticism or apology; but it is believed that the circumstances under which the writing was performed should be recorded, for the benefit of inquiring historians of the future. Much could be said on the other side; of the invaluable assistance furnished by all who were consulted; of the extremely useful reports and descriptions prepared by University and other contractors on many phases of the work; and of the untiring, conscientious efforts of the writers themselves.

1-4. How the History is Built.

a. Mechanical Division - The Manhattan District History is composed of eight books, each comprised of one or more separate volumes. Some books are composed of a considerable number of volumes (Book I has 12 Volumes, numbered from 1 to 14, with Volumes 2 and 3 omitted); some books have extending "Supplements" and some have "Top Secret" Appendices or "Top Secret" Supplements. Nearly every volume has appendices of its own, and in many cases these appendices, which may attain considerable bulk, are bound in separate covers.

The appendices contain maps, drawings, diagrams, photographs, letters, reports, and other documents which are particularly useful or desirable as supplements to the main text and are not too bulky for inclusion. One appendix in nearly every volume consists of references only, giving the title and file location of other material — usually original source material — so that any reader who has access to the files may, if he wishes, pursue any subject more intensively.

At the beginning of each volume there is a brief Foreword, stating pertinent facts which may have influenced the writing, the date to which the history covered by the volume has been written, and usually (if this date is earlier than 31 December 1946) whether or not the volume is accompanied by an extending appendix. The Foreword is followed by a Table of Contents and then by a Summary of the volume (usually less than 20% of the length of the main text). At the end of each volume there is usually an Index, and, in the more technical volumes, one of the appendices is a Glossary of unusual words or expressions not found in an ordinary dictionary.

It is not necessary to give here a mere list of the titles of all the books and volumes — such a list appears in Appendix B of this volume — nor is it necessary to summarize here the books or the volumes — such summaries appear in each volume — but, to orient the reader and to help him to find specific subjects, the general framework of the History will be described and (in a following section) a few words about each of the succeeding books, will be presented.

      b. <u>General Framework</u> - The division of the History into books has been made, in general, by major, physically separated, functional parts of the project: Book II covers the Gas Diffusion Project (at Oak Ridge); Book III, the P-9, or Heavy Water, Project (at various munitions plants in the U. S. and at an industrial plant in Canada); Book IV, the Pile Project (at Hanford, Washington); Book V, the Electro-magnetic Project (at Oak Ridge); Book VI, The Liquid Thermal Diffusion Project (at Oak Ridge); and Book VIII, the Los Alamos Project (principally at Los Alamos). Book VII is devoted to "Feed Materials, Special Procurement and Geographical Exploration "for the Manhattan Project as a whole, while Book I, of which this is a part, covers: those subjects which are in general pertinent to the project as a whole or pertinent to two or more of the subjects treated in other books; and certain subjects which are termed "Auxiliary Activities", consisting principally of activities undertaken after the cessation of hostilities, during the period of conversion from war-time to peace-time operation, and of activities which are in special categories and, for one reason or another, do not belong in any other volume.

      In dealing with the subsidiary projects (of the Manhattan Project as a whole), in the books concerned with the various different processes of manufacture or production (such as the Gas Diffusion Project, and the Electromagnetic Project), the framework of each book, typically, is composed of volumes having the following self-explanatory titles: Vol. 1, General Features; Vol. 2, Research; Vol. 3, Design;

Vol. 4, Construction; Vol. 5, Operation; and, usually, a Supplement concerned with production. Some variations occur in special cases; for example: Book III, the P-9 Project, is contained entirely in one volume, in which the first five sections bear titles corresponding to the volume titles listed above, while a sixth section bears the title: "Organization and Personnel"; Book IV, the Pile Project, has an additional volume inserted as number 4: "Land Acquisition"; and Book VI (the Liquid Thermal Diffusion Project), which is also contained in a single volume, is composed of sections entitled: 1. Introduction; 2. Research; 3. Design and Construction; 4. Description of Plant; 5. Operations; 6. Organization and Personnel.

Book VIII, the Los Alamos Project, differs markedly in its framework from the other books devoted to production projects, primarily because of the different manner in which this project was set up and administered, and because of the numerous ramifications of the activities with which this project was concerned outside of Los Alamos. This book has three volumes: Vol. 1, General; Vol. 2, Technical; and Vol. 3, Auxiliary Activities.

1-5. What the History Contains.

a. Book I, General - All except one (Volume 4) of the twelve volumes of this book cover subjects which are generally pertinent to the project as a whole, or to two or more of the subjects treated in other books (as mentioned above). These subjects are made fairly obvious by the titles of their volumes. Volume 4 covers the "Auxiliary Activities" previously described.

Vol. 1, "General" (this volume), presents the introduction to the History, the Mission of the Manhattan District, the Authorizations for the work of the Manhattan District as a whole, the Total

Costs, and, in appendices, a Combined Table of Contents, Combined Indices

(of the names of persons and organizations), ~~and~~ a Map showing the locations

of Manhattan District Installations, and a General Index of the whole History.

Vol. 2, and Vol. 3 are omitted.

Vol. 4, "Auxiliary Activities", is divided into 14 chapters and

one Supplement, the titles of which describe them sufficiently for pre-

sent purposes:

Chap. 1, Legislative Contacts of Manhattan District.

Chap. 2, Foundation of the National Laboratories.

Chap. 3, Program for Production and Distribution of Radio-
isotopes.

Chap. 4, Research and Development of Atomic Energy for Power.

Chap. 5, Declassification and Distribution of Project
information.

Chap. 6, Investigation of the After Effects of the Bombing
in Japan.

Chap. 7, Contributions of Representatives of the Manhattan
District to the Discussions and Proposals for International Control.

Chap. 8, Press Releases.

Chap. 9, Assistance on the Canadian Pile Project.

Chap. 10, The Oak Ridge Institute of Nuclear Studies.

Chap. 11, Ames Project (Iowa State College).

Chap. 12, Activities of the National Bureau of Standards.

Chap. 13, Preparation and Publication of the Smyth Report.

Chap. 14, Investigations of Miscellaneous Processes of
Separation of Uranium Isotopes.

Top Secret Supplement, Storage Project.

Vol. 5, "Fiscal Procedures", describes the general plan of fiscal management of the Manhattan District and the considerations on which it was based. It deals with estimates, budgets, audits, finance, disbursing and cost accounting.

Vol. 6, "Insurance Program", describes the various types of insurance coverage plans which were developed and instituted for the joint protection of the United States, the contractors and all the employees, under the unusual and unknown conditions of hazard which the project produced.

Vol. 7, "Medical Program", describes, in non-technical language, how the Manhattan District guarded the health and safety of the employees (except at Los Alamos, which is covered in Book VIII). It deals with the many unique problems which arose from the previously unknown processes of the project; pioneering research work in the effects on the human body of new types of radiation and radioactive materials; industrial hygiene; and clinical and public health programs. This volume describes, in connection with the radioactive and the chemical hazards which were encountered, the determination of safe tolerance levels, and the development of methods and instruments for monitoring and measuring exposure.

Vol. 8, "Personnel", describes how the Manhattan District procured and maintained the manpower required in all phases of its operations (except at Los Alamos, which is covered only in part in this volume). It deals with: recruiting; conservation and utilization of personnel; labor relations; wage and salary administration; selective service; and procurement and administration of military and naval personnel.

Vol. 9, "Priorities Program", describes how the "Washington Liaison Office" and other District offices expedited the procurement of materials and manufactured products required in the District's operations (exclusive of Feed Materials and Special Procurement, which are covered in Book VII). It tells how the District cooperated with the War Production Board and other agencies, to insure the least possible interference with other government and civilian activities, without misuse of the high priority rating to which the District was entitled.

Vol. 10, "Land Acquisition, CEW", describes the procedures followed in the acquisition of the site, of more than 58,000 acres of land, for the Clinton Engineer Works and the town of Oak Ridge, in Roane and Anderson counties, near Knoxville, Tennessee. This volume deals with direct purchase and with acquisition by condemnation, and it includes data showing progress of purchases, acreages, appraisal values and amounts paid. The history of the selection of the site will be found in Volume 12 (of Book I). This site serves the Gas Diffusion Project, the Electromagnetic Project, the Liquid Thermal Diffusion Project, and the Clinton Laboratories (covered in Book IV), as well as the town of Oak Ridge, and for this reason this volume is placed in Book I.

Vol. 11, "Safety Program", describes the procedures which were instituted and administered by the District and its contractors in the interest of safety and accident prevention. It presents statistics of employee-hours of occupational exposure and measures the safety record of the District by frequency, severity and fatality rates. (From this volume also Los Alamos has been excluded.)

Vol. 12, "Clinton Engineer Works, Central Facilities", describes the selection of the entire site (the acquisition of which is described in Vol. 10), and the planning, installation or construction, administration and operation of those facilities which serve all the plants which are located thereon, as well as the town of Oak Ridge. In a part entitled "Town of Oak Ridge", it deals with: town planning; construction; operations; housing; commercial facilities; schools; medical facilities and services; and social and welfare facilities and services. In a part headed "Area Facilities", this volume deals with: the electrical system; the water supply system; the sewerage system; communications; roads, streets, walks and bridges; railroad system; passenger transportation; and security. Guard and police forces, fire protection, and administration of justice are among the subjects included under security. As with Vol. 10, "Land Acquisition, CEW", these Central Facilities serve a number of the subsidiary parts of the Manhattan Project as a whole (which are covered in other books) and therefore this volume also appears in Book I.

Vol. 13, "Patents", describes the development and operation of the patents program, first under OSRD and then under the Manhattan District, primarily for the purpose of protecting the Government's interest in inventions produced during the prosecution of the work, and of securing the maximum possible control of atomic energy by means of patents. The patents program also included: the execution of decisions, in the field of foreign relations, made by the Combined Policy Committee of Great Britain, Canada, and the United States; the filing of foreign applications; and the purchase of patent rights from private inventors.

Under executive order, all Government patent rights relating to the project are held in the custody of the Director of OSRD, Dr. Vannevar Bush. This volume deals also with the Standard patent clauses inserted in the Manhattan District contracts, and describes how the patent activities were carried out without danger of violating the security requirements of the project.

Vol. 14, "Intelligence and Security", describes the development and operation of the program whereby secrecy and protection were maintained throughout the Manhattan Project, to prevent espionage, sabotage, damage, interference or other harm which might endanger or delay the project. It deals with: investigation of potential subversives; personnel security; plant protection; control of visitors; protection of shipment (of both documents and material); safeguarding military information; and the means whereby the security of Clinton Engineer Works, at Oak Ridge, as a whole, was protected. (The local security programs at Hanford and at Los Alamos are described in Book IV and Book VIII respectively.)

"Foreign Intelligence Supplement No. 1", to Volume 14, describes the activities of the "ALSOS Mission" in Europe, following closely on the heels of the victorious armies, and the activities of a special mission to Japan, which, after the surrender, investigated the progress which the Japanese had made in nuclear research. "Foreign Intelligence Supplement No. 2" describes the activities for detection of enemy use of radioactivity during the invasion of Normandy; and "... No. 3" describes export control activities.

b. Book II, Gaseous Diffusion (K-25) Project, deals with one of the processes for the separation of uranium isotopes, whereby

fissionable material, concentrated U_{235}, is obtained, for incorporation
in the U_{235} type of atomic bomb. This process is based on the principle
that the rate of diffusion of a gas through an ideal porous barrier is
inversely proportional to the square root of its molecular weight. Thus,
when uranium hexafluoride is diffused through a porous barrier, the
lighter isotope, U_{235}, diffuses more rapidly than the heavier isotope,
U_{238}, and this causes separation. By the use of many successive diffusion stages the desired concentration and quantity of U_{235} are obtained.
The research work for this part of the Manhattan Project was carried on
principally at Columbia University and the large-scale production plant
(K-25, and its extension K-27) was constructed and operated at the
Clinton Engineer Works, Oak Ridge, Tennessee. It was planned by the
Kellex Corporation, built by J. A. Jones Construction Co., Inc., and
operated by Carbide and Carbon Chemicals Corporation. During its earlier
history, this plant supplied partially enriched feed for the Electromagnetic plant; near the end (in December 1946) it was operated so
as to produce independently the product required for final processing
and incorporation in the bomb (at Los Alamos).

c. Book III, P-9 Project, deals with manufacture of "heavy
water", or deuterium oxide. Although this material was not actually
used in the production of the atomic bombs, it was thought during the
early stages of the Manhattan Project that it might be necessary for
use as a moderator in the manufacture of plutonium (described in Book
IV). The production of heavy water was regarded as particularly
important when it became known that the Germans were undertaking large
scale production, as it was thought at that time that the enemy might

CONFIDENTIAL/RD   SECRET

be nearer than the Allies were to production of atomic bombs. Research
and experimentation have been carried on, and are continuing , to
determine the possibilities of the use of heavy water and it may be
that they will result in improved methods of plutonium manufacture
in the future. (Reference should be made to Book I, Volume 4, Chapter
9.)

Book III covers the various processes for producing heavy
water which were considered, and, in detail, the two which were used:
(1) the "hydrogen gas" process, whereby the heavy water was obtained
from natural water and hydrogen, by catalytic exchange reaction, using
hydrogen generated electrolytically for the manufacture of ammonia; and
(2) the "water distillation" process, whereby the heavy water was obtained
principally by separation from up-flowing steam and transfer to down-
flowing water, in a series of columns or stages, the final concentration
being obtained electrolytically. The research work for this part of the
project was carried on, for OSRD and for the Manhattan District, at a
number of places, by a number of contractors, including: Standard Oil
Development Co., Columbia University, Consolidated Mining & Smelting
Co. of Canada, Ltd., and E. I. du Pont de Nemour & Co. A production
plant, using the hydrogen gas process, was set up in Trail, B. C.,
Canada, as an adjunct to the plant of the Consolidated Mining and
Smelting Company of Canada, Ltd., where hydrogen was being manufactured;
it was built and operated by that company. Plants for production by
the water distillation process were installed and operated by the
duPont Company at three existing Government Ordnance Works in the
United States: the Morgantown, the Wabash River, and the Alabama

CONFIDENTIAL/RD   SECRET   RESTRICTED DATA
ATOMIC ENERGY ACT 1946

Ordnance Works. All these matters are described in Book III.

d. Book IV, Pile Project, X-10, deals with the manufacture of plutonium, another fissionable material, required for incorporation in the plutonium type of atomic bomb. It covers the research and development work carried on at the Metallurgical Laboratory at the University of Chicago and at the Clinton Laboratories at Oak Ridge, Tennessee, and the design, construction and operation of the large scale production plants at the Hanford Engineer Works, on the Columbia River, near Pasco, Washington. In these plants, by means of a controlled self-sustaining "chain reaction", in a water-cooled uranium-graphite "pile", uranium 238 is transmuted by fission first to neptunium and thence to plutonium, and then the plutonium is chemically separated from the residual uranium and the undesirable impurities known as fission products.

The book describes, as necessary background information, the pioneer research work of scientists in foreign countries and in the United States, and the further work done under the OSRD, as well as the final research and development accomplished under the Manhattan District. A notable mile-stone in the history of the project – one which deserves to be remembered as a mile-stone in the history of the world – was the completion and successful operation of the first chain-reacting pile, at the Metallurgical Laboratory on 2 December 1942. Then for the first time it was proved definitely that a nuclear chain reaction could be produced by man.

This book covers also all the central facilities and services, and those subjects which pertain generally, not only to the production plant but also to the community which was built up for the resident workers on the project - the town of Richland. Among the general items covered are: the selection of the site; the acquisition of the site, comprising a total of about 631 square miles; power transmission and distribution; communications; roads and railroads; housing; commercial facilities; service facilities; utilities; inter-city transportation facilities; intelligence and security; and law enforcement.

This book tells how E. I. duPont de Nemour and Company designed and built this part of the Manhattan Project, and operated it from the beginning until 1 September 1946; and how on that date the operation was taken over by the General Electric Company.

e. Book V, Electromagnetic (Y-12) Project, deals with another of the processes for the separation of uranium isotopes, whereby concentrated U_{235} (the same fissionable material which is produced in the Gas Diffusion Project) is obtained, for incorporation in the U_{235} type of atomic bomb. This process, sometimes called the mass-spectrographic method, is based on the principle that when ionized molecules of a gaseous compound are projected into a magnetic field they will move in semi-circular paths of radii proportional to their momenta; and the light ions will move in smaller semicircles than the heavy. Thus, a beam of particles of natural uranium, fired from an ion source into a magnetic field, will split into two semicircular beams, the outer one consisting of particles of mass 238 and the inner one of particles of mass 235. A receiver, placed at the point of maximum separation, will collect the U_{235} particles and the U_{238} particles in separate pockets. By the use of a sufficient number of separate batch units and stages, the

desired quantity and concentration can be obtained. The research work
for this part of the Manhattan Project was carried on principally at the
University of California Radiation Laboratory, and the large-scale pro-
duction plant was built and operated at the Clinton Engineer Works, Oak
Ridge, Tennessee. Design and construction were carried out by Stone and
Webster Engineering Corporation and the plant was operated by Tennessee
Eastman Corporation.

One of the unusual features of this project was the use of
a large quantity of silver - about 14,700 tons - as a substitute for
copper, as an electrical conductor in the electromagnets in the production
plant. Copper was one of the most critical of war-time materials, and,
by procuring silver from the U. S. Treasury, copper was saved for other
important war uses. Volume 4 of this book describes the procurement,
safeguarding and fabricating of, and the accounting for, this silver.

f. Book VI, Liquid Thermal Diffusion, (S-50), Project, deals
with a third process for the separation of uranium isotopes, by which concentrated U_{235} was obtained, in this case to serve as feed material for
further enrichment in the Electromagnetic plant, pending the completion
and operation of the Gas Diffusion Plant. This process is based on the
principles of: (1) thermal diffusion, which causes higher concentration
of the lighter component of some liquid mixtures near the hotter of two
surfaces; and, (2) convection, which carries the lighter material near
the hotter wall upward, thus causing a difference in concentration verti-
cally. Thus, when uranium hexafluoride was passed through a column with
a steam-heated inner core and a water-cooled outer shell (circulated by

means of a convection loop) a product enriched in U_{235} was obtained at
the top of the column, leaving depleted material at the bottom. By
recirculation and the operation of a sufficient number of columns the
desired concentration and quantity of U_{235} were obtained. The research
work for this part of the Manhattan Project was carried on principally
by the Naval Research Laboratory (following earlier research at the
Carnegie Institution of Washington). The large-scale production plant
was built and operated at the Clinton Engineer Works, Oak Ridge,
Tennessee. It was planned and constructed by the H. K. Ferguson Company,
and was operated by Fercleve Corporation, a subsidiary of that company,
until September 1945, when it was shut down and placed in stand-by
condition.

g. Book VII, Feed Materials, Special Procurement and
Geographical Exploration, deals, in Volume 1, with the procurement of
the basic raw materials containing uranium, and their processing to the
condition required for the feed to various production plants; this
volume also describes the procurement of various special materials
required for the operations of the Manhattan District. The procurement
of all special materials is not covered by this book, but only those
which were handled by the District organization which was charged with
the procurement of the Feed Materials: the so-called Madison Square
Area, located in New York City. Volume 2, of this book covers the
world-wide geographical exploration (both bibliographical and in the
field) carried on, principally under the direction of the Murray Hill
Area (also in New York City), to determine the location and extent of
occurrences of the basic raw materials.

h. Book VIII, Los Alamos Project (Y), deals with all the activities which were carried on at the Los Alamos Laboratories, Los Alamos, New Mexico, including principally: research, design, final processing, construction, assembly and testing of the bombs themselves; site selection and land acquisition; construction, administration and operation of the community, with its housing, utilities, commercial facilities and services; and construction, administration and operation of the so-called technical area and of various outlying, "satellite", sites. This book also tells of numerous other activities which were carried on at other places scattered through the United States, all of which may be regarded as auxiliary to the critical major work at Los Alamos, some of them of very great importance. The technical work at Los Alamos was carried on under contract with the University of California, directed by Dr. J. R. Oppenheimer of that University; the general administration of the community, which was operated as a military post, and all construction work were under the local direction of an Army officer - the Commanding Officer. Major construction work was performed principally, in the earlier days, by M. M. Sundt Construction Company, and then, through successive large expansions, by Robert E. McKee. The Architect-Engineer for most of the design work was W. C. Kruger.

The non-scientific part of the history of the Los Alamos Project is in Volume 1 of Book VIII; the technical part of its history is in Volume 2; and the auxiliary activities are described, in various chapters, in Volume 3.

Of the auxiliary activities, particular mention should be made of: the Los Angeles Procurement Office, which handled most of the diverse procurement required for all the Los Alamos activities; and the Camel Project, which covered some complex research, testing and production work and served as an emergency reserve in case of catastrophe at Los Alamos. Other important auxiliary subjects described in Volume 3 include: Activities of Ohio State Cryogenic Laboratory; Dayton Project; Navy Participation; Sandia; Boron; Operation Crossroads (in which personnel of the Manhattan District took an important part); and various other miscellaneous activities.

The real climax of the Los Alamos Project - indeed, the real climax of the whole of the Manhattan Project - is described in the "Technical" volume of this book. This was the so-called Trinity Test, the first man-made atomic bomb explosion of all time, which took place in the desert at Alamogordo Air Base, New Mexico, on 16 July 1945. Not until that test was successfully made could anyone connected with the enterprise be sure that the project would produce any concrete results.

The official climax of the Manhattan Project - the official culmination of the years of effort of thousands of people and of the expenditure of about two billions of dollars - was the combat bombing of Japan, at Hiroshima and Nagasaki, in August 1945. This also is described in the "Technical" Volume (Vol. 2) of Book VIII, in a chapter headed: "Project Alberta".

SECTION 2 - THE MISSION OF THE MANHATTAN DISTRICT

2-1. <u>The District's Mission, in Simplest Terms</u> - Reduced to its simplest terms, the mission of the Manhattan District was to develop and manufacture atomic bombs for combat use, at the earliest possible date, and ahead of the enemy, in order to help the United States and its Allies to bring World War II to a successful conclusion with the least delay and the greatest possible conservation of life.

2-2. <u>Justification</u> - The world's history during the first few weeks following the dropping of the first combat bomb on Hiroshima on 6 August 1945 proved conclusively not only that this mission of the Manhattan District was successfully accomplished but also that this mission - with its accompanying expenditure of about two billions of dollars - was completely justified.

a. One phase of this justification has been described impressively by Major General L. R. Groves, in a public address given in June 1946, in the following words:

"....While the mission was to develop the atomic bomb, the real objective was to save the lives of thousands of America's best men - by shortening the war.

"The war was not won by the atomic bomb. It was won by our fighting men, backed up by the best of weapons provided for them by the industrial might of the nation. The Japanese had lost the war long before the Hiroshima bombing, but the people did not know it, and their leaders did not admit it. Both people and leaders knew it and admitted it within a few hours after they learned of Hiroshima.

"I believe the historians of the future will be unanimous in the conclusion that the atomic bomb ended the war in a more abrupt and sudden manner than any war has been ended since the day of that other great surprise weapon, the Trojan Horse.

"Let us look for a moment at the timetable of last summer. On the 16th of July, the first atomic bomb test occurred in New Mexico. News of the successful results were rushed to President Truman, then in Potsdam, and in a few days, - on the 26th, to be exact - the Potsdam surrender ultimatum was sent to Japan, warning her that unless she surrendered she would be destroyed.

"On the 29th, Radio Tokyo broadcast that the Japanese Premier scorned the Allied warning as 'unworthy of public notice.' Eight days later, Hiroshima was hit by the first atomic bomb. We told the government and the people of Japan what had hit them. That was August 6, 1945. Two days later, Russia hurriedly entered the war against Japan. The next day, the second atomic bomb fell on Nagasaki. On the tenth of August, the Swiss Charge d'Affaires in Washington sent a note to our Secretary of State, stating that the Japanese government was ready to accept the terms of the Potsdam Ultimatum. On the 14th, after almost four years of unremitting war and eight days after the first atomic bomb was dropped, the Japanese government unconditionally surrendered.

"The men scheduled to enter Japan the hard way - across the beaches - and their mothers and fathers, wives and children, soon realized that they had had an infinite stake in the two billion dollar program which had been such a closely guarded secret for more than three years . . ."

b. Another and more terrifying phase of the justification of the mission of the Manhattan District was the threatening possibility that the enemies of the United States might produce and use the atomic bomb first. In 1939, when President Roosevelt was first apprised of the potentialities of uranium fission, this possibility of enemy precedence was impressed upon him – and he took the courageous steps that ultimately set the Manhattan District in motion. It was surmised at that time that: (1) the production of a uranium fission bomb might prove to be possible; (2) the Nazis were secretly accumulating large stocks of uranium ore; (3) heavy water, in quantity, could be used effectively in the production of fissionable uranium material; and (4) the Nazis were manufacturing heavy water in quantity. For those who knew these things, both in Great Britain and in the United States, the dread conclusion was inescapable, that the Nazis might soon produce atomic bombs.

c. It remains for future history to pass judgment upon still another phase of the justification of the mission of the Manhattan District: the position of the United States as the sole possessor of this greatest military weapon of all time and the advantage this gives her, toward remaining ahead of other nations in its development and operation, and toward seeking international cooperation for its control.

2-3. Indirect Justification – Future history must also pass judgment upon those results of the work of the Manhattan District which lie beyond the scope of its original mission. These are the non-military results, which are not concerned with national or international defense. The possibilities of the contributions of these results to science, industry and medicine – to civilization as a whole – cannot yet be

estimated, but their part in the indirect justification of the
District's original mission can nevertheless be dimly imagined.

SECTION 3 - AUTHORIZATIONS AND LINES OF AUTHORITY

3-1. <u>General</u> - The Manhattan Project was an ultra-secret war-time project, and as such it was unthinkable that it could be authorized in the ordinary way, by direct Congressional enactment and by direct Congressional appropriation. It was one project above nearly all others which demanded the exercise of the broad general war-time powers conferred upon the President and the War Department. If the Congress had passed an act specifically authorizing the project, the unavoidable publicity attendant upon such action would have made secrecy impossible from the beginning.

The authority under which the work of the Manhattan District was carried out stemmed to a considerable extent from the authority under which much of the early research work was begun, under the auspices of the National Defense Research Committee and the Office of Scientific Research and Development; and even before the NDRC and the OSRD were authorized to undertake early parts of the work, some official actions were taken, and these also form an important part of the record.

The stories of authorizations and lines of authority are inextricably linked together and their history is therefore best presented by giving a brief résumé, step by step, of the genesis and the purpose of each important organization through which authority was exercised, from the President down. This will be succeeded by a description of the Congressional acts by which general war-time powers were conferred upon the President and the Secretary of War, and of the orders by which these powers were redelegated by them.

Even the bare recital of the organizational means whereby authority was exercised impresses upon the reader unmistakably the extreme care taken by those in responsible positions not to abuse the powers under which they acted. Here was a project which, if successful, might change the whole course of history; but, correspondingly, if it failed, all those concerned might be condemned by the whole nation. The stakes were so high that extreme risks were thoroughly justified. Risks were taken -- onerous risks; but they were always carefully calculated risks. Before any important step was taken, especially until the "all-out" effort was finally under way, the best available advice was sought and obtained -- from statesmen, scientists, engineers and military men. This procedure accounts to some extent for the comparatively large number of committees, groups, etc., which were involved in the early stages of the work.

3-2. <u>Earliest Government Interest</u> - The first government interest and support of research in nuclear physics dates from 11 October 1939, when Dr. Alexander Sachs of New York, armed with a letter from Dr. Albert Einstein, called on President Franklin D. Roosevelt and explained to him the desirability of government encouragement of work in this field. Two nuclear physicists, Dr. Leo Szilard and Dr. E. Wigner, had previously conferred with Dr. Einstein and Dr. Sachs, and the letter which Dr. Sachs brought to the President resulted from these conferences. (App.A-1,pp.7ff; App.A-2,par. 3.4ff.)

3-3. <u>Advisory Committee on Uranium</u> - The President then appointed the "Advisory Committee on Uranium", composed of Dr. L. J. Briggs, Chairman, Colonel K. F. Adamson of the Army Ordnance Department, and

RESTRICTED DATA
ATOMIC ENERGY ACT 1946

Commander G. C. Hoover of the Navy Bureau of Ordnance. Meetings held by this Committee were attended by leading scientists in the field of nuclear physics, and a report of the committee to the President dated 1 November 1939 contained specific recommendations, mentioning the possibilities of both atomic power and an atomic bomb. (App. A-2, par. 3.4ff; App. A-3, p.423.)

3-4. **First Transfer of Funds** - Shortly after, the first small bit of public funds, $6,000 - the beginning that grew to an ultimate expenditure of nearly $2,000,000,000 - was transferred from the Army and Navy, for the purchase of materials recommended by the committee. (App. A-2, par. 3.6, this transfer was reported by Dr. Briggs in a memorandum dated 20 February 1940; App. A-3, p.423.)

3-5. **Timely Research Results at Columbia** - Meanwhile, the research work at Columbia University produced some encouraging results, which were very opportune in influencing the further progress of the project at this critical time, in the spring of 1940. (App. A-2, par. 3.7.)

3-6. **National Defense Research Committee** - In June 1940, the National Defense Research Committee was organized and the President issued instructions that the Uranium Committee, (of which Dr. Briggs was chairman) should become a subcommittee of that organization, reporting to its chairman, Dr. Vannevar Bush. The NDRC was a part of the Office of Scientific Research and Development, which was also headed by Dr. Bush, Director. (Dr. J. B. Conant of Harvard University later became chairman of NDRC.) The status of the Uranium subcommittee remained substantially unchanged until the summer of 1941, when it was enlarged and became known as the Uranium Section or the S-1 Section of

NDRC. (App. A-2, par. 3.9, 3.14; App. A-3, p. 423.)

3-7. First Contracts - The first contract under this organization was let to Columbia University, Contract NDCcr-32, signed 6 November 1940, effective from 1 November 1940 to 1 November 1941. This contract called for the expenditure of $40,000 for work which had been recommended by a special advisory committee which Dr. Briggs had called together on 15 June 1940. Other contracts were made, with Universities and others, until, by November 1941, sixteen projects, totalling about $300,000, had been approved. (App. A-2, par. 3.10-3.12; App. A-3, p. 424.)

3-8. National Academy Reviewing Committee - At the request of Dr. Briggs and Dr. Bush, Dr. F. B. Jewett, President of the National Academy of Sciences, appointed in the spring of 1941 a special committee, to review the problem. The members of this committee were: Dr. A. H. Compton, Chairman; Drs. W. D. Coolidge, E. O. Lawrence, J. C. Slater, J. H. Van Vleck and B. Gherardi. Later (before the committee's second report was made), Mr. O. E. Buckley of the Bell Telephone Laboratories and Mr. L. W. Chubb of the Westinghouse Electrical and Manufacturing Company, were added to the Committee; and still later (before the committee's third report), Drs. W. K. Lewis, R. S. Mulliken and G. B. Kistiakowsky were appointed as additional members. This committee's first report, in May 1941, formed the basis for approval by the NDRC, on 18 July 1941, of an appropriation of $267,000 for the project; at the same time, it was indicated that much larger expenditures would probably be necessary. In this and in a subsequent report the National Academy Reviewing Committee urged that the project be pushed more vigorously; in a third report, dated 6 November 1941, emphasis was placed, for the first time, on the "possibilities of an explosive fission reaction with U_{235}", and on the probable decisive importance of the project in the current war. A particularly significant paragraph from this report should be quoted:

> "If all possible effort is spent on the program, one might however expect fission bombs to be available in significant quantity within three or four years."

This statement was made exactly three years and nine months before Hiroshima. (App. A-2, par. 3.15, 3.16, 4.48, 4.49; App. A-3, pp. 424, 425.)

3-9. Top Policy Group - Following the first two reports of the National Academy Reviewing Committee, and after receiving optimistic reports from the British, Dr. Bush discussed the whole project with President Roosevelt. The President agreed that the whole program should be enlarged and reorganized, and arranged that determinations of general policy should be made by the "Top Policy Group", composed of: the President, the Vice-President, the Secretary of War, the Chief of Staff, Dr. Bush and Dr. J. B. Conant. (App. A-2, par. 3.22; App. A-3, p. 427.)

3-10. Reorganization.

a. OSRD S-1 Section - In November 1941, after the third report of the National Academy Reviewing Committee was issued, Dr. Bush decided that the time had come for an "all out" effort to develop atomic bombs for use in the current war and that this would require a reorganization of the existing NDRC Uranium Section; accordingly he and Dr. Conant arranged for that section to be divorced from NDRC and become a new OSRD S-1 Section. This was announced to the members by Dr. Conant on 6 December 1941. With this reorganization, the direction of the uranium projects was, in effect, placed in the hands of: Dr. Bush and Dr. Conant; Dr. Briggs, Chairman; Drs. A. H. Compton, E. O. Lawrence, and H. C. Urey, program chiefs; and Dr. E. V. Murphree, chairman of a separately organized Planning Board. (App.A-2,par.5.2-5.8; App.A-3, p.428.)

b. OSRD S-1 Executive Committee - In May 1942, Dr. Bush terminated the OSRD S-1 Section and replaced it with the OSRD S-1 Executive Committee, consisting of: Dr. Conant, Chairman, and Drs. Briggs, Compton, Lawrence, Murphree and Urey, thus, in a sense, confirming officially the direction of the work which already existed unofficially. (App.A-2, par.5.17; App.4.)

3-11. Activation of the Manhattan District.

a. Report of March 1942 - In a report to the President on 9 March 1942, Dr. Bush first suggested that the Army be brought into the project, with the recommendation that, during the summer of 1942, the Army should be authorized to construct full-scale plants. (App.A-2, par.5.13.)

b. Report of 13 June 1942 - a Presidential Directive - The March report of 9 March was followed by the all-important report of 13 June 1942, of Drs. Bush and Conant, which was approved by Vice President Henry A. Wallace, Secretary of War Henry L. Stimson, and Chief of Staff General George C. Marshall; on 17 June 1942, this report was also approved by the President.

Because the President's initials on this report of 13 June

1942 set off a chain of events which resulted, immediately, in the activation of what became known as the Manhattan District, and, ultimately, in the successful attainment of the atomic bomb, description of the contents of this report in some detail is justified. (App.A-5; App.A-2,par.5.21ff; App.A-3,pp.435,436.)

The report was divided into four principal parts, as follows:

(1) <u>Status of the Program</u> - The scientists in charge of the various phases of the program were unanimous in believing that the production of atomic bombs was possible. Specifically:

(a) A mass of a few kilograms of U_{235} or plutonium would create an explosion; the energy released would be equivalent to several thousand tons of TNT; and the explosion could be controlled to occur at the desired instant.

(b) There were four methods of preparing the fissionable materials, all of which appeared to be feasible; but it was not possible to state definitely that any one of them was superior to the others.

(c) Production plants of considerable size could be designed and built.

(d) With adequate funds and priorities, full-scale plant operation could be started soon enough to be of military significance in the current war.

(2) <u>Recommendations of the Program Chiefs and the Planning Board</u> - The program chiefs and the chairman of the Planning Board, all members of the S-1 Executive Committee, recommended the

following program, to be undertaken at the earliest possible date:

(a) Construction of a centrifuge plant for substantial yield of U_{235}, to be scheduled for completion in January 1944.

(b) Construction of a gas diffusion pilot plant for the separation of U_{235} and complete design of a full-scale gas diffusion plant.

(c) Construction of an electromagnetic plant for substantial yield of U_{235}, to be scheduled for completion late in 1943.

(d) Construction of an atomic power (pile) plant for substantial production of plutonium early in 1944.

(e) Construction of plants to yield substantial quantities of heavy water, beginning 1 May 1943, as auxiliaries to the atomic power plant.

(f) Continuation of fundamental studies in physics and chemistry at an accelerated pace.

(3) <u>Comments on the Recommendations</u> - The recommendations outlined above had been reviewed by Dr. Bush and Dr. Conant, and by General W. D. Styer, who had been instructed by General Marshall to follow the progress of the program. The comments of these three, which were incorporated in the report, may be summarized as follows:

(a) If four separate methods all appeared to a highly competent scientific group to be capable of successful application, it seemed certain that the end result could be attained by the enemy, provided he had sufficient time.

(b) The proposed program obviously could not be carried out rapidly without interfering with other important matters.

(c) It would be unsafe, at that time, in view of the pioneering nature of the entire effort, to concentrate on only one means of obtaining the result.

(d) The best procedure, therefore, appeared to be to proceed at once with those phases of the program which would interfere least with other important war activities; and to proceed with the other phases after questions of interference had been determined by further study.

(4) Recommendations of Dr. Bush and Dr. Conant - A fourth part of the report (which has been referred to also as a "covering memorandum" - Baxter, p.435) contained the following recommendations of Dr. Bush and Dr. Conant:

(a) That research and development continue under OSRD contracts, and that the sum of $31,000,000 be made available to OSRD for financing the following: pilot plants for the centrifuge and gas diffusion processes; research and development on the electromagnetic process; a full-scale electromagnetic plant; a heavy water project; and miscellaneous research; and that a contingent fund of approximately $5,000,000 be made available for the fiscal year 1943, the allocation of this fund and the direction of the work to be in the hands of an executive committee appointed by the Director of OSRD.

(b) That construction of the plants and the development of the power plant project be put in charge of a qualified Army Officer designated by the Chief of Engineers and reporting to him; and that this officer be assisted by Drs. E. V. Murphree, A. H. Compton, and L. W. Chubb, or others equally qualified, on a full-time basis, the last-named two being in charge of the research and engineering aspects of the power plant development, respectively.

(c) That $54,000,000 be made available to the Chief of Engineers for the fiscal year 1943, but that, in order to avoid delay and suspension of the work, the Chief of Engineers be authorized to make further expenditures or over-obligate any funds under his control, with the understanding that he would be reimbursed for justified expenditures at a later date.

(d) That contracts be let by the Chief of Engineers, without delay, after consultation with members of the OSRD Planning Board, for the detailed design of all plants indicated above.

(e) That a site, or sites, be selected and acquired, with proper consideration of power requirements; that an immediate start be made on the construction of the necessary fencing, housing, utilities, and other facilities required; and that on one of these sites the development of the power pilot plants be carried out, under such arrangement as the officer in charge should determine after consultation with the S-1 Executive Committee of OSRD.

(f) That coordination of research, development and construction be assured by frequent meetings of the S-1 Executive Committee of the OSRD with suitable officers designated by the Chief of Engineers, and that these meetings should submit progress reports as required.

(g) That a special committee, or committees, on the military uses of the material to be produced be arranged through the Joint Committee on New Weapons and Equipment, of the Joint Chiefs of Staff; and that such committee, or committees, be assigned charge of all research and development on this aspect of the project.

(h) That, as soon as the preparation of the detailed plans and designs had progressed sufficiently, the highest priority be assigned to the plant or plants which at that time showed the most promise of success and indicated the least serious effect, in demands for critical materials, upon other urgent programs.

(i) That the remaining plants be assigned priority and preference ratings after evaluation of bills of material in comparison with requirements for other urgent programs; and that their construction be executed accordingly.

(j) That the greatest secrecy be exercised in connection with the project, particularly with respect to its purpose, the raw materials used to develop the final product, the final product, and the manufacturing processes involved.

(k) That the plants to be constructed be camouflaged under suitable names; and their purpose be announced in a similar camouflaged manner.

c.   Departures from the Presidential Directive - With the President's approval of this report it became a directive to the Army and to the OSRD. This directive set the broad outlines of action and policy for the intensive prosecution of the whole project and although changes in some details of these outlines became necessary from time to time, the directive as a whole remained nevertheless an influence and a guide throughout the work of the three years following. Among the changes which developed later were: the abandonment of the centrifuge process, because of failure of early experiments; the addition of

RESTRICTED DATA
ATOMIC ENERGY ACT 1945

the liquid thermal diffusion process, as an auxiliary, because of the
excellent and promising results obtained from experiments in this process; the by-passing, in effect, of the pilot plants, by undertaking
construction of full-scale plants before the pilot plants could be
placed in operation, in order to save every possible day of time; the
withdrawal of the OSRD as the supervising and administering agency for
research and development; and the placing of the entire project under
the planning direction of a new "Military Policy Committee". (App. A-2
p.5.25; App.A-8.)

That part of the directive which dealt with the future concentration of effort on the most promising plant or plants was the cause
of considerable uncertainty. There was a great deal to be said in favor
of "backing one horse" instead of all, primarily because of the threatened
shortages of manpower and critical materials. At a meeting on 26 August
1942, Dr. Conant expressed the opinion that it was then too soon to
make such a decision, and it was fortunate indeed that decision was
made at that time to continue to "back all five horses"; If one
method had been chosen at that time, it would have been the electromagnetic, and although the electromagnetic plant contributed vitally
to the early success of the project, its difficulties increased as the

---

*Footnote: The "five horses" mentioned by Dr. Baxter in "Scientists
Against Time" ("the brief official history of the OSRD") were: (1)
the Electromagnetic process; (2) the gas diffusion process; (3) the
centrifuge process; (4) the uranium-graphite pile process; (5) the
uranium-heavy-water pile process.

work progressed and success could probably not have been attained in time with this plant alone. (App. A-3, p.436.) As a matter of fact, except for the changes mentioned above, no horse dropped out until the race was really over, when, in September 1945, the liquid thermal diffusion plant was shut down and placed in stand-by condition; then, in December 1946, the electromagnetic plant (the operations of which had been materially curtailed somewhat earlier) was taken out of production of atomic bomb material altogether, and, except for its chemical treatment facilities, was operated only in part, for auxiliary non-military purposes.

d. Army Organization.

(1) District Engineer - On 18 June 1942, the day after the President affixed his approval to the above described report of Drs. Bush and Conant and made it a Presidential directive, General Styer advised Colonel James C. Marshall, Corps of Engineers, that he had been selected by the Chief of Engineers to form a new Engineer District, to carry on special work assigned to it (App. A-2, par.5.23). Colonel Marshall was then District Engineer of the Syracuse District, and, as this District had recently completed the major part of its war construction program, he was able to bring with him a small nucleus of key personnel without any delay. Colonel Marshall served as District Engineer of the new District until he was succeeded in August 1943 by his former deputy, Colonel, K. D. Nichols. Colonel Nichols served as District Engineer until after 31 December 1946. (Colonel Nichols was advanced to Brigadier General on 22 January 1946, but, under the post-combat readjustment of grades of high-ranking officers, he reverted to a

colonelcy on 30 June 1946. Some time after the termination of the period covered by the Manhattan District History, on 27 April 1948, he became Major General. He is generally referred to throughout this History as "Colonel" Nichols, the rank he held during most of his active work as District Engineer. With regard to the lower ranking officers of the District, whose grades were in some cases subject to considerable variation by promotions, the usual – but not invariable – policy has been followed in this history of referring to each one by the rank he held at the time concerned in the reference.) Colonel Marshall and Colonel Nichols brought with them, from the Syracuse District, Miss Virginia Olsson, and she served as Secretary to the District Engineer throughout the history of the Manhattan District.

(2) <u>Manhattan District</u> – Although the new District was technically in existence from the date of the selection of its District Engineer, 18 June 1942, its official activation was effective as of 16 August 1942, under authority of General Orders No. 33 of the Office of the Chief of Engineers, dated 13 August 1942. (App. A-9.) By this order the new District was designated, probably for the first time officially, The Manhattan District. Colonel Marshall was officially assigned as District Engineer, and Colonel (then Lt. Col.) Nichols as Assistant District Engineer, Manhattan District, by Special Orders No. 177 of the Office of the Chief of Engineers, dated 13 August 1942 (App. A-10).

In conformity with the camouflage policy recommended by Drs. Bush and Conant, the work of the District was designated the "DSM Project" (Development of Substitute Materials).

(3) <u>Major General L. R. Groves</u> – In a memorandum dated 17 September 1942, from the Commanding General, Services of Supply

(later, Army Service Forces), to the Chief of Engineers, Major General
(then Colonel) L. R. Groves, by direction of the Secretary of War,
was relieved of his assignment as Deputy Chief of Construction, in
the Office of the Chief of Engineers, and placed in complete charge
of all Army activities relating to the atomic bomb project (App. A-6).
General Groves was to report directly to the Military Policy Committee
(which was about to be appointed) and to the Chief of Staff, the
Secretary of War and the President. Thus the project was removed from
the direct supervision of either the Chief of Engineers or the Commanding
General of the SOS (or its successor, the ASF). General Groves was
instructed to: (1) operate in conjunction with the Construction Division
of the Office of the Chief of Engineers and with other facilities of
the Corps of Engineers; (2) take immediate steps to arrange for necessary
priorities; (3) arrange for a working committee on application of the
product; (4) arrange for procurement of a site, or sites, and transfer
activities thereto; (5) initiate the preparation of construction bills
of material; (6) draw up plans for the organization, construction,
operation and security of the project, and, after approval, put them
into effect. (Colonel L. R. Groves became Brigadier General on
23 September 1942; and Major General on 9 March 1944. Some time after
the termination of the period covered by the Manhattan District History,
on 24 January 1948, he became Lieutenant General. He is generally
referred to throughout this History as "Major General", or, briefly,
"General", Groves.) General Groves brought with him, from the Office
of the Chief of Engineers, Mrs. Jean O'Leary, and she served as his
secretary throughout the history of the Manhattan District.

3-12. Military Policy Committee - On 23 September 1942, a conference was held by those who were designated by the President to determine the general policies of the project. Those present were: Secretary of War Henry L. Stimson, Army Chief of Staff General George C. Marshall, Dr. V. Bush, Dr. J. B. Conant, General (then Major General) Brehon Somervell, Lieutenant General (then Major General) W. D. Styer, and Major General (then Brigadier General) L. R. Groves. (Vice-President Henry A. Wallace was unable to attend.) At this meeting a Military Policy Committee was appointed, consisting of Dr. Bush as Chairman, with Dr. Conant as his alternate, General Styer, and Rear Admiral W. R. Purnell. General Groves was named to sit with this committee and to carry out the policies which were determined. The duties of this committee were to plan military policies relating to materials, research and development, production, strategy, and tactics, and to submit progress reports to the top policy group designated by the President. The appointment of the Military Policy Committee was approved by the Joint New Weapons Committee, established by the Joint Chiefs of Staff and consisting of Dr. Bush, Admiral Purnell and Brigadier General R. G. Moses. (App. A-7, -8.)

3-13. Combined Policy Committee - In the beginning, international relations between the United States and Great Britain and Canada in connection with the atomic bomb project were handled by the President of the United States and the Prime Ministers of Great Britain and Canada. In August 1943, however, it became apparent that an international committee would be desirable, and this led to the establishment of the Combined Policy Committee, with the following membership: for the United States, Secretary of War Henry L. Stimson, Dr. Vannevar Bush, and Dr. James B. Conant; for the United Kingdom, Field Marshal Sir John Dill and Colonel J. J. Llewellin; and for Canada, Mr. C. D. Howe. Colonel Llewellin was replaced by Sir Ronald I. Campbell in December 1943 and the latter, in turn, was replaced by the Earl of Halifax. The late Field Marshal Sir John Dill was replaced by Field Marshall Sir Henry Maitland Wilson early in 1945. The scientific adviser of the United States members was Dr. Richard C. Tolman; of the British members, Sir James Chadwick; and of the Canadian member, Dean C. J. Mackenzie. (App. A-17.)

An important function of the Combined Policy Committee was the provision for interchange of information between the nations. It was arranged that full interchange would be maintained in the field of scientific research and development, but that in matters of design, construction and operation of large scale plants information would be exchanged only when such exchange would hasten the completion of weapons for use in the current war. (App. A-17.)

3-14. _Continuation of Authority_ - The organizational set-up described herein above continued through the further progress of the project and until the mission of the Manhattan District had been accomplished. To recapitulate, its major administrative components were:

a. President Franklin D. Roosevelt and, after his death in April 1945, President Harry S. Truman.

b. The Top Policy Group: The President, the Vice President, the Secretary of War, the Chief of Staff, Dr. Bush and Dr. Conant.

c. The Military Policy Committee: Dr. Bush, Chairman, and Dr. Conant, his alternate; General Styer; Admiral Purnell; and General Groves, Executive Officer.

d. The S-1 Executive Committee of the OSRD: Dr. Conant, Chairman, and Drs. Briggs, Compton, Lawrence, Urey, and Murphree.

e. The Manhattan District: General Groves, Commanding General; Colonel Marshall, succeeded by Colonel Nichols, District Engineer.

The further organizational components of the Manhattan District and the authorizations under which they acted, and the contractual relations between the District and the many contractors who served on the project, are described in subsequent parts of the History.

3-15. <u>General War Power Authorizations</u> - As previously stated, the President and the Secretary of War were granted wide war-time powers by the Congress, and it was under these that they authorized the work of the Manhattan District. Under these, also, they redelegated authority to others.

a. <u>Public Law No. 703 - 76th Congress, 3rd Session, approved 2 July 1940</u> - This law was enacted by the Congress to expedite and facilitate the strengthening of the National Defense. It authorizes the Secretary of War to provide for the necessary construction, rehabilitation, conversion, installation and operation of plants and buildings, for the development, manufacturing and storage of military equipment and supplies, and for shelter.

b. <u>Public Law No. 354 - 77th Congress, 1st Session, approved 18 December 1941 - "First War Powers Act"</u> - This law provides that the President may authorize any department or agency of the Government, exercising functions in connection with the war effort, to enter into contracts and modifications without regard to previous laws pertaining to contracting; except that the cost plus a percentage contract may not be used, and the laws pertaining to limitations of profits must not be violated.

c. Public Law No. 507 - 77th Congress, 2nd Session, approved 27 March 1942 - "Second War Powers Act". - This law (in Title II) authorizes the Secretary of War to acquire by purchase, donation, transfer, or condemnation, any real property, temporary use thereof, or other interest therein, together with any personal property located thereon or used therewith, that shall be deemed necessary for military or other war purposes.

d. Public Law No. 580 - 77th Congress, 2nd Session, approved 5 June 1942, providing amendments to Public Law No. 703, described above - This law provides for sundry matters affecting the military establishment and authorizes the Secretary of War to employ architectural and engineering technical and professional firms and individuals to design and supervise the construction of additional War Department facilities, when, in his opinion, the existing facilities are inadequate.

e. Executive Order No. 9001, dated 27 December 1941, by which, with subsequent amendments, the President delegated certain of his powers under the "First War Powers Act" to the Secretary of War (App. A-11).

f. Order of the Secretary of War, dated 30 December 1941, by which the Secretary delegated to the Under Secretary of War powers conferred upon him by Executive Order No. 9001 described above (App. A-12).

3-16. Additional Specific Authorizations - In addition to the specific authorizations described in preceding paragraphs (including particularly the Presidential Directive described in some detail in paragraph 3-11,b), the following are worthy of mention:

a. <u>Directive on Extraordinary Security Measures</u> - On 29 June 1943 the President informed General Groves that because of its great significance to the nation the atomic bomb project was to be more drastically guarded than other highly secret war developments and pointed out that, as General Groves knew, he (the President) had given instructions that every precaution be taken to insure the security of the project. (App. A-13.)

b. <u>Delegation of Authority by the Under Secretary of War</u> - On 17 April 1944 the Under Secretary of War delegated to Major General Groves, effective as of 1 September 1942, the authority to exercise contractual powers under Executive Order No. 9001, "in connection with the work assigned to and coming within the jurisdiction of the Manhattan District, U. S. Engineer Office" (App. A-14).

c. <u>Delegation of Authority by General Groves</u> - On 10 June 1944, General Groves delegated to the District Engineer, Manhattan District, effective as of 1 September 1942, authority to enter into contracts for the Manhattan District, subject, however, to written approval for any contract exceeding $5,000,000 in price (App. A-15).

d. <u>Authority to Settle Claims against the United States</u> - On 17 June 1944, the Under Secretary of War redelegated, to the District Engineer, Manhattan District, the authority vested in him to consider, ascertain, adjust, determine, settle and pay claims against the United States which arose from the activities of the Manhattan District, subject to the purview and limitations of Public Law No. 112, 78th Congress, and to appeal to the Secretary of War (App. A-16).

3-17. <u>Appropriations</u> - The appropriations from which the funds for the Manhattan District were allotted may be found in Book I,

Volume 5, Fiscal Procedures.

SECTION 4 - TOTAL COST

4-1. "Two Billion Dollars." It has been commonly stated that the total cost of the Manhattan Project was about "two billion dollars". This was the figure cited by President Truman in his announcement of 6 August 1945, when he said, "We have spent two billion dollars on the greatest scientific gamble in history - and won." Actually, the total expenditures up to that date were somewhat less than two billion dollars, but the commitments already made, and the continuation of the operations of the Manhattan District for the further period of seventeen months, through 31 December 1946, produced a final total figure of somewhat more than two billion dollars. As a round figure, "about two billion dollars" remains substantially correct, as the approximate measure of the total cost of the whole of the Manhattan Project. It is even more nearly accurate to say that the amount spent on the atomic bomb (through 31 December 1946) was about two billion dollars. The final overall total costs of the Manhattan District included the costs of various activities which were concerned with peace time applications of atomic energy, and not directly with the development and production of the bomb. Those expenditures which were not directly concerned with the bomb were responsible for some of that part of the total cost which exceeded two billion dollars.

4-2. Accounting for the United States Atomic Energy Commission. When, in accordance with the terms of the Atomic Energy Act of 1946, the Manhattan District prepared, at the direction of the President, to transfer to the Atomic Energy Commission all its facilities, equipment, materials, etc., an accounting was prepared, on the basis of the cost of the entire

work. Because of the size, complexity and widely scattered locations
of the assets of the Manhattan District, this accounting took many months
to complete, after the termination of the control of the District on
31 December 1946. The results were embodied in the "Manhattan District
Project Cost Summary for the period ending June 30, 1947". Although the
period extended to the end of the Fiscal Year 1947, the expenditures
enumerated in the summary included only those made from Manhattan District
funds - the so-called "Military" expenditures. Actually, in closing up
the work as of 31 December 1946, some expenditures were made from these
funds as late as February 1947, and these also are reflected in the
Summary, together with all credits, adjustments, and "contra accounts".
The final figure for total costs in the Summary represents the total
overall cost of the Manhattan Project as accurately as it is possible
to determine. This figure is $2,163,393,503 and it appears in the
Summary, in the report of the status of funds as of 30 June 1947, as
follows:

Total Allotments or Allocations	$2,163,393,983
Total Costs	2,163,393,503
Balance of Allotments or Allocations	$ 480

The figure for "Total Allotments or Allocations" quoted from the Summary
above does not agree with the total reported in the History, Book I,
Volume 5, Fiscal Procedures, wherein it is stated (page 2.6): "As of
December 31, 1946, the net allotment to the Manhattan Project totaled
two-billion-three-hundred and eleven million dollars"; but this is because
of revocations and because of transfers to the AEC, made subsequent to
the preparation of the figures given in that volume.

The volume on Fiscal Procedures states (page 6.2) that the cost vouchers processed by the District Cost Section through 31 December 1946 resulted in "gross costs of approximately $2,200,000,000", and this agrees reasonably closely with the figure quoted from the Summary. As stated further in the volume on Fiscal Procedures, however, certain costs were excluded in recording the gross costs, as follows:

"(1) Allotments by the Chief of Engineers to other agencies." for work performed for the District.

"(2) Transportation Corps (open allotments).

"(3) Finance Department (pay, allowances, and travel of the Army).

"(4) Miscellaneous procurements and depot issues of other agencies not charged to Manhattan Funds or vouchered as free issues.

"(5) Expenditures by OSRD prior to the establishment of the Manhattan District."

4-3 Breakdown of Costs. The Manhattan Project Cost Summary for the period ending 30 June 1947, to which reference has been made above, gives a breakdown of the total costs, which can be edited to show, in a somewhat limited manner, the total costs of the principal parts of the Project. These total costs do not in all cases agree with those which are cited in various other books and volumes of this History, for a number of reasons, including:

a. All items of costs and credits were not yet available when some of the books were written.

b. Classifications of costs were not always made on the same basis in compiling the separate sets of cost figures.

c. Distribution of cost items, particularly such as might apply to two or more of the separate installations, was dependent upon individual judgment and may have been made in accordance with different theories and different methods of estimating.

The following table (somewhat simplified by the combination of items of the same title or description) is quoted from the Summary:

Item		Amount
Y-12 and Special Accounts		$553,296,580
K-25		604,123,610
S-50		16,277,101
X-10		479,399,561
Semi-works		25,116,166
F-9		27,906,509
Los Alamos		182,241,148
C.E.W. and Central Facilities		160,337,721
Medical and General		5,968,313
Special Materials & Procurement		213,351
General		37,019,666
Government Overhead		38,344,574
General Service Contracts	Credit	190,327
Post War Programs		20,713,516
Total - Feature Accounts		$2,150,767,489
Adjustment - Contra Accounts	Credit	29,635,986
Net Cost		$2,121,131,503
Insurance and Welfare		42,262,000
Total Costs Chargeable to M.D. Funds		$2,163,393,503

APPENDIX A

REFERENCES IN SECTION 3

APPENDIX A

REFERENCES IN SECTION 3

1. Testimony of Dr. Alexander Sachs, 27 November 1945; Hearings before the Special Committee on Atomic Energy, United States Senate, 79th Congress, First Session, Part 1 (Printed GPO), pages 7 ff. Manhattan District History files, Washington, D. C.

2. "Atomic Energy for Military Purposes - The Official Report on the Development of the Atomic Bomb, under the Auspices of the United States Government, 1940-1945"; by Henry DeWolf Smyth, Chairman, Department of Physics, Princeton University, Consultant, Manhattan District, U. S. Engineers; Princeton University Press, 1945 (with detailed index compiled in the Office of the Commanding General, Manhattan District). Manhattan District History files, Washington, D. C.

3. "Scientists Against Time"; by James Phinney Baxter 3rd, President, Williams College; Little, Brown and Company, 1946. ("This is the brief official history of the Office of Scientific Research and Development" - Vannevar Bush, in Foreword, 21 January 1946). Manhattan District History files, Washington, D. C.

4. Minutes of OSRD S-1 Executive Committee. M.D. Class. Files.

5. Letter, 17 June 1942, from Dr. V. Bush to President Roosevelt, inclosing letter dated 13 June 1942 from Drs. V. Bush and J. B. Conant to Vice President H. A. Wallace, Secretary of War Henry L. Stimson, and Army Chief of Staff George C. Marshall. Class. Files of Major Gen. L. R. Groves, Washington, D. C.

6. Memorandum of 17 September 1942, from Commanding General, Services of Supply, to the Chief of Engineers. Class. Files of Major Gen. L. R. Groves, Washington, D. C.

7. Memorandum "A" dated 23 September 1942 signed by: Drs. V. Bush and J. B. Conant; Vice President Henry A. Wallace; Secretary of War Henry L. Stimson; Army Chief of Staff George C. Marshall. This was attached to a memorandum of transmittal, dated 1 October 1942, from Dr. V. Bush to Major Gen. (then Brig.Gen.) L. R. Groves. Class. Files of Major Gen. L. R. Groves, Washington, D. C.

8. Minutes of Meetings of Military Policy Committee. Class. Files of Major Gen. L. R. Groves, Washington, D. C.

9. General Orders No. 33, 13 August 1942, Office Chief of Engineers. Copy in Manhattan District History files, Washington, D. C. (Early Draft, Bk. I, Vol. 1).

10. Special Orders No. 177, 13 August 1942, Office Chief of Engineers. Copy in Manhattan District History files, Washington, D. C. (Early Draft of Bk. I, Vol. 1).

11. Executive Order No. 9001, 27 December 1941, as Amended: Regulations under the First War Powers Act, 1941. Copy in Manhattan District History files, Washington, D. C. (Early Draft, Bk. I, Vol. 1).

12. Memorandum, 30 December 1941, from the Secretary of War to the Under Secretary of War. Copy in Manhattan District History files, Washington, D. C. (Early Draft, Bk. I, Vol. 1).

13. Letter, 29 June 1943, from President Roosevelt to Major General (then Brigadier General) L. R. Groves. Copy in Bk. I, Vol. 14, App. A-1.

14. Memorandum from the Under Secretary of War to Major General L. R. Groves, dated 17 April 1944. Photostat in Manhattan District History files, Washington, D.C. (Early Draft, Bk. I, Vol. 1).

15. Memorandum from Major General L. R. Groves to the District Engineer, Manhattan District, dated 10 June 1944. Photostat in Manhattan District History files, Washington, D. C. (Early Draft, Bk. I, Vol. 1).

16. Memorandum from the Under Secretary of War to the District Engineer, Manhattan District, dated 17 June 1944. Copy in Manhattan District History files, Washington, D. C. (Early draft, Bk. I, Vol. 1.).

17. Statement by the Secretary of War, 6 August 1945. Copy in Department of State Publication 2702. (See also Bk. I, Vol. 4, Chap. 8).

APPENDIX B

COMBINED TABLE OF CONTENTS

APPENDIX B

COMBINED TABLE OF CONTENTS

OF

MANHATTAN DISTRICT HISTORY

B.1

(Book I, Vol. 4, Chap. 4, Cont'd)

(Book I, Vol. 4, Cont'd)

(Book I, Vol. 6, Cont'd)

Vol. 7, Medical Program (1 cover)

Vol. 8, Personnel (1 cover)

Vol. 9, Priorities Program (1 cover)

(Book I, Vol. 9, Cont'd)

(Book I, Vol.12, Cont'd)

Part A - General Introduction
    Section 1 - Description
            2 - Site Selection and Preparation
            3 - Basic Considerations

Part B - Town of Oak Ridge
    Section 4 - Planning and Design
            5 - Construction
            6 - Operations
            7 - Housing
            8 - Commercial Facilities
            9 - School System
           10 - Medical Facilities and Services
           11 - Social and Welfare Facilities and Services

Part C - Area Facilities
    Section 12 - Electrical System
           13 - Water Supply System
           14 - Sewerage System
           15 - Communications
           16 - Roads, Streets, Walks, and Bridges
           17 - Railroad System
           18 - Passenger Transportation Service
           19 - Security System
    Appendix A - File References   )
             B - Maps   )   (separate cover)
             C - Graphs, Charts and Tables)
             D - Photographs and Plans (separate cover)

Vol. 13, Patents (1 cover)

    Section 1 - Introduction
            2 - Policy
            3 - Applications for Patents
            4 - Contract Administration
            5 - Records
            6 - Organization
            7 - Co-operating Agencies
    Appendix A - Documents
             B - Patent Division Manual

Vol. 14, Intelligence and Security (5 covers)

    Section 1 - Introduction
            2 - Counter-Intelligence
            3 - Personnel and Company Clearance

(Book I, Vol. 14, Cont'd)

                4 - Plant Protection
                5 - Shipment Security
                6 - Security of Information
                7 - Organization
        Appendix A - Letters and Organization Chart
                 B - Memoranda and Intelligence Bulletins
                 C - Miscellaneous Forms and Manuals
                 D - References
                 E - Copies of Newspaper Clippings
                 F - Personnel of Intelligence and Security Division
Supplement (31 December 1945 to 31 December 1946) (separate cover)
Top Secret Appendix to Supplement (separate cover)
Foreign Intelligence Supplement No. 1 (separate cover)
        Section 1 - Introduction
                2 - Establishment of ALSOS Missions
                3 - Italian Investigation
                4 - Western and Central European Investigation
                5 - Investigation of Nuclear Research in Japan
                6 - Organization and Personnel
        Appendices A, B   and 3
Foreign Intelligence Supplements Nos. 2   (separate cover)

BOOK II, GASEOUS DIFFUSION (K-25) PROJECT

Vol. 1, General Features (1 cover)

Section 1 - Introduction
        1-1, Purpose
        1-2, Scope
        1-3, Authorization
        1-4, Administration

Section 2 - Basic Theory
        2-1, Nuclear Fission
        2-2, Concentration of Uranium 235
        2-3, Application of Theory

Section 3 - Planning of the Project
        3-1, Beginning of the Project
        3-2, Research and Development
        3-3, Design and Engineering
        3-4, Construction
        3-5, Operation

Section 4 - Description of the Project
        4-1, Description of Site
        4-2, Description of Facilities
        4-3, Size of the Project

CONFIDENTIAL/RD   SECRET

RESTRICTED DATA
ATOMIC ENERGY ACT 1946

(Book II, Vol. 1, Cont'd)

DOE
b(3)

RESTRICTED DATA
ATOMIC ENERGY ACT 1946

Book II, Vol. 3, Cont'd)

DELETED   DELETED   DELETED   DOE b(3)

Section 12 - Power Plant Design

(Book II, Vol. 3, Section 18, Cont'd)

Vol. 4, Construction (1 cover)

Section 1 - Introduction

Section 2 - Contractual Arrangements

Section 3 - Construction of Facilities

RESTRICTED DATA
ATOMIC ENERGY ACT 1946

(Book II, Vol. 4, Cont'd)

DELETED   DELETED   DELETED

D015
6(3)

(Book II, Vol. 5, Section 8, Cont'd)

BOOK III. THE P-9 PROJECT (HEAVY WATER)   (1 volume, 1 cover)

BOOK IV. PILE PROJECT. X-10

(Book IV, Vol. 1, Cont'd)

(Book IV, Vol. 2, Part I, Cont'd)

(Book IV, Vol 2, Part II, Section 4, Cont'd)

(Book IV, Vol. 3, Section 1, Cont'd)

1-5, Design Considerations

CONFIDENTIAL/RD SECRET

(Book IV, Vol. 5, Section 3, Cont'd)

(Book IV, Vol. 5, Cont'd)

(Book IV, Vol. 6, Section 3, Cont'd)

3-6, Third Specification Conference

(Book IV, Vol. 6, Section 9, Cont'd)

(Book IV, Vol. 6, Section 14, Cont'd)

BOOK V. ELECTROMAGNETIC (Y-12) PROJECT

(Book V, Vol. 1, Section 1, Cont'd)

(Book V, Vol. 2, Cont'd)

CONFIDENTIAL/RD ~~SECRET~~ ~~RESTRICTED INFORMATION~~

(Book V, Vol. 3, Section 2, Cont'd)

RESTRICTED DATA
~~ATOMIC ENERGY ACT~~

(Book V, Vol. 4, Cont'd)

Book V, Vol. 4, Section 7, Cont'd)

(Book V, Vol. 5, Cont'd)

(Book V, Vol. 6, Section 7, Cont'd)

BOOK VI, LIQUID THERMAL DIFFUSION (S-50) PROJECT (1 volume, 2 covers)

(Book VI, Section 4, Cont'd)

BOOK VII. FEED MATERIALS, SPECIAL PROCUREMENT AND GEOGRAPHICAL EXPLORATION

Vol. 1. Feed Materials and Special Procurement (TOP SECRET, 1 cover)

Part A - General Features

(Book VII, Vol. 1, Part A, Section 1, Cont'd)

(Book VII, Vol. 1, Part C, Cont'd)

(Book VII, Vol. 1, Appendix, Cont'd)

        H - Miscellaneous Materials for P-9
        I - Miscellaneous Materials for X-10
        J - Procurement for Site Y
        K - Beryllium Procurement

BOOK VIII, LOS ALAMOS PROJECT (Y)

~~CONFIDENTIAL/RD~~ ~~SECRET~~

~~CONFIDENTIAL/RD~~ B.45 ~~SECRET~~ RESTRICTED DATA
ATOMIC ENERGY ACT 1946

(Book VIII, Vol. 2, Chapter IV, Cont'd)

4.54, The Water Boiler

(Book VIII, Vol. 2, Chapter IX, Cont'd)

(Book VIII, Vol. 2, Chapter XIV, Cont'd)

(Book VIII, Vol. 2, Supplement, Chapter IV, Cont'd)

(Book VIII, Vol. 3, Chapter 1,   Section 2, Cont'd)

(Book VIII, Vol. 3, Cont'd)

(Book VIII, Vol. 3, Chapter 3, Part II, Section 3, Cont'd)

APPENDIX C

COMBINED INDICES OF MANHATTAN DISTRICT HISTORY

C(1) Index of Names of Persons

C(2) Index of Names of Agencies, Industrial
Organizations, Universities, etc.

C(1) <u>INDEX OF NAMES OF PERSONS</u>

<u>Notes:</u>

1.   References, generally, are made to Book numbers and Volume numbers, as: "II-1", "V-3" etc. The references to Volume 4 of Book I and to Volume 3 of Book VIII also specify the Chapter numbers thus: "I-4:5" (Bk.I,Vol.4,Chap.5), "VIII-3:2" (Bk.VIII,Vol.3,Chap.2),etc. The references to Volume 2 of Book IV also specify the Part number (Part I or Part II), thus: "IV-2I" or "IV-2II". The references to Book III and Book VI contain no Volume or Chapter numbers. Individual references are separated by commas. Where a name is mentioned in two or more Volumes of the same Book, the number of the Book is given with the first reference only; thus, "I-1,4:3,VIII-1,2,3:6" refers to the following five parts of the History: Volume 1 of Book I; Chapter 3 of Volume 4 of Book I; Volume 1 of Book VIII; Volume 2 of Book VIII; and Chapter 6 of Volume 3 of Book VIII.

2.   A name in this Index will usually be found in the Index of the Chapter, Volume or Book referred to, and there further reference designates the page number (or paragraph number), except that in some cases the name may be found only directly in the text or in an appendix.

3.   Because the writers of some parts of the History (notably Bk.VIII,Vol.2) have followed the policy of omitting the titles of civilians, such as "Dr." or "Prof.", the names do not always appear in this Index with their proper titles. For officers of the Army and Navy, the highest rank by which each is mentioned in the History is usually (but not always) given in this Index.

4.   No references to Top Secret Appendices or Top Secret Supplements are included in this Index.

5.   Some names, not included in the main index, have been inserted as "Addenda", on p. C(1) 24.

Abelson, Dr.P.H., I-4:12, IV-1, 2I, VI
Acher, H.M., VIII-1
Acheson, Dean, I-4:7
Ackart, E.G., III, IV-1, 2II
Acker, F.M., IV-6
Ackerman, Maj.J.C., VIII-2
Adam, Lt.Francis O., Jr., I-14
Adams, J.P., VIII-1
Adamson, Col.K.F., I-1
Aebersold, Dr.E.C., I-4:3
Agnew, H., VIII-2
Agnor, Capt.G.L., VI
Akeley, Maj.W.G., VII-1
Akers, Dr.W.A., II-2
Albrechtson, E.B., IV-6
Alexander, E.C., I-4:6
Alexander, Dr.P.P., VII-1
Alexander, R., VII-1
Allen, Lt.Geo.S., I-14
Allen, H.S., VIII-2
Allen, Dr.J.G., I-7
Allen, Lt.J.H., I-4:6
Allinson, J.J., II-4
Allis, Maj.W.P., I-14
Allison, Dr.S.K., I-4:4, 4:9, 4:12,
   4:13, IV-2I, VIII-2
Alspaugh, P.L., II-5
Alter, H.W., I-4:7
Alvarez, Dr.L.W., I-4:7, VIII-2
Amaldi, Dr., I-14
Ambros, Lt.Wayne A., I-14
Ames, Capt.Kenneth, VIII-1
Amis, Robt.T., I-8
Anderson, Lt.Andrew J., Jr., I-14
Anderson, Dr.C.D., VIII-3:2
Anderson, Ens.D.L., VIII-2, 3:8
Anderson, Capt.G.W., Jr., USN, VIII-3:8
Anderson, H.L., VIII-2
Anderson, Capt.J.D., II-3
Anderson, Maj.N.D., I-5
Anderson, Cmdr.Roland A., I-13
Anderson, Lt.Wilbur S., I-14, II-5
Anderson, Lt.Wm.A., Jr., I-14
Andre, Gaston, I-14
Andrews, Rep.W.C., VIII-3:8
Anicetti, Dr.R.J., VII-1

Antes, Col.D.E., I-4:1, I-5, VIII-3:1
Appen, H.V., II-4
Appley, Lawrence A., I-9
Archer, Maj.N.R., II-3, 5
Ardinger (Fassett and), IV-4
Argersinger, R.E., V-3
Argue, Fred, V-3
Armistead, F.C., II-5
Arneson, R.Gordon, I-4:7
Arnold, Gen.H.H., VIII-3:8
Arnold, Dean Samuel T., I-8, VIII-2, 3:9
Ashbridge, Lt.Col.Whitney, VIII-1, 2
Ashworth, Cmdr.R.L., VIII-2, 3:8
Aston, Dr.F.W., I-4:14, II-2
Aton, Lt.Thos.J., I-14
Attlee, Prime Minister C.R., I-4:7
Aug, G.C., I-12
Ayers, A.N., VIII-2
Ayers, John A., I-4:11
Aylor, Lt.J.H., I-14
Azevedo, Lt.Daniel, I-14

Babcock, D.F., IV-6
Bacher, Dr.R.F., I-4:2, 4:5, 4:7, VIII-2,
   3:9
Badger, R.M., II-2
Baerwind, Dr., I-14
Bagge, Dr.Errich, I-14
Bagley, J.G., II-5
Bain, Geo.W., VII-2
Bainbridge, Dr.K.T., I-4:3, VIII-2
Baird, Maj.Douglas O., I-13
Baird, Lt.F.M., I-14
Baker, Charlie, VIII-3:9
Baker, C.P., VIII-2
Baker, E.P., V-6
Baker, F.E., IV-6
Baker, Jas., VIII-2
Baker, Dr.Lynn, I-7
Baker, L.G., VIII-3:1
Baker, Nicholas, VIII-2
Baker, R.D., VIII-2
Bale, Dr.W.F., I-7
Balke, C.C., VIII-2

Bamer,L.G.,I-11
Banikiotes,G.,VIII-3:7
Baranowski,F.P.,II-5
Barge,Lt.R.I.,I-14
Barkley,Sen.Alben W.,I-4:1
Barnard,Chester I.,I-4:7
Barnes,Capt.Guy E.,I-14
Barnes,Lt.Philip,VIII-2
Barnett,Capt.H.L.,I-4:6
Barnett,Capt.M.J.,III,VIII-3:4
Barnett,Lt.O.C.,I-14
Barnett,S.C.,II-5
Barrett,Capt.Martin K.,I-5
Barrett,W.F.,II-5
Barrish,Capt.J.S.,IV-5
Barron,Dr.E.S.G.,I-7
Barschall,H.H.,VIII-2
Bartholomeu,C.Y.,VIII-3:7
Bartley,Dr.Walter,I-4:13
Bartnett,R.T.,I-11
Baruch,B.M.,I-4:7
Baskin,Lt.J.R.,I-14
Bassett,Capt.L.G.,VII-1
Batson,Maj.R.T.,I-4:5
Baut,Lt.H.S.,I-14
Baxter,Dr.J.P.,I-4:6
Baxter,Maj.Samuel,I-12
Bayer,Lawrence,VIII-3:8
Beahan,Capt.K.K.,VIII-2
Beams,Dr.J.W.,I-4:12,4:14
Beaudin,Milton,I-9
Beck,C.K.,II-5
Beck,E.E.,VIII-2
Becker,Dr.August,I-14
Beckwith,Maj.M.K.,II-5
Becquerel,H.,IV-1
Bederson,T/5 B.,VIII-2
Beeler,Col.Geo.W.,I-4:2,4:7,IV-6,
  VII-1
Beerman,Lt.F.L.,I-14
Begg,Capt.,VIII-3:9
Behling,H.A.,IV-5
Belcher,Maj.Fred.H.,I-12,II-3,5,V-3
Belcher,Lt.P.F.,I-14
Belgerod,Capt.H.,VIII-1
Belina,Frank,VII-2

Bell,Daniel W.,I-9,V-4
Bell,E.W.,I-5
Belmore,F.M.,VII-1
Benbow,Maj.H.S.,I-4:9,VII-1
Benedict,Dr.M.,I-4:7,II-5
Bennett,Lt.Albert,I-14
Berg,Lt.C.M.,VIII-1
Bergelin,Capt.O.,VII-1
Berkie,Dr.,I-14
Best,S.J.,VIII-3:4
Bethe,H.A.,VIII-2
Betts,Col.A.W.,VIII-2
Bezman,Dr.I.I.,VIII-3:3
Bibbs,W.E.,VIII-2
Bibby,Harry,I-6
Bice,R.A.,VIII-2,3:8
Bigelson,Lt.Gerald,VIII-1
Biles,D.,VII-1
Birch,Lt.Cmdr.A.F.,VIII-2
Bishop,H.R.,I-6
Blackistone,F.D.,I-5
Blair,Col.R.C.,I-10,12
Blair,T.,VIII-3:7
Blakely,F.F.,II-5
Blanchard,Lloyd E.,VIII-2
Blandy,Vice Adm.W.H.P.,VIII-3:8
Blauvelt,Lt.R.W.,I-14
Bliss,Dr.L.A.,I-4:7,II-5
Bloch,Maj.Edw.J.,I-4:2,12
Bloch,F.,VIII-2
Block,Maj.L.R.,I-14
Block,Lt.Melvin A.,I-4:6
Block,Seymour,I-4:6
Bloom,Dr.Wm.,I-7
Blum,Capt.Bernard,I-7
Bochette,Lt.J.W.,I-14
Boche,Dr.Robt.,I-7
Bogart,Maj.D.C.,I-5
Bogash,R.,VIII-3:7
Boggs,E.M.,VIII-3:9
Bohden,Capt.V.L.,I-9
Bohr,Aage,VIII-2
Bohr,Niels,IV-1,2I,VIII-2
Bolstad,M.,VIII-2
Bolton,Lt.F.H.,I-14
Boltzmann,L.,VIII-2

Bondrake, L.D., VIII-2

Bonesteel, Col. C.H., VIII-3:8

Boorse, H.A., II-2

Booth, Dr. E.T., II-2,5

Bopp, Dr., I-14

Borden, Brig. Gen. W.A., VIII-3:8

Borst, Dr. L.B., I-4:2

Bothe, Prof. Walther, I-14

Bowelle, B., IV-3

Bowen, Rear Adm. H.G., I-4:12, VI,
    VIII-3:8

Bowen, Dr. I.S., VIII-3:2

Bower, H.C., II-5

Bowes, Watson, VIII-1

Bowman, Prof. H.R., I-4:6

Boyd, Dr. Geo., I-7

Bradbury, Dr. N.E., I-4:5, II-5, VIII-2,
    3:5, 3:8

Bradway, F.W., I-6

Brady, Lt. Max, I-14

Branch, R.T., V-3

Brannon, Capt. J.H., II-3

Bransome, E.D., VII-1

Bratt, E.C., VIII-3:4

Bray, Ralph, VIII-3:9

Brazier, B.E., VIII-1,2

Breit, Dr. G., I-4:12

Brennand, J.R., VIII-1

Brenner, Richard, I-9

Brereton, Gen. L.H., VIII-3:8

Bretscher, E., VIII-2

Brewer, Dr. A. Keith, I-4:14

Bridges, Sen. Styles, I-4:1

Bridgman, W.P., VIII-2

Briggs, Dr. L.J., I-1, 4:12, 4:14, II-3

Bright, W., VIII-2

Brin, T/Sgt. R., VIII-2

Briney, R.M., VII-1

Brinkman, Maj. E.A., III, VII-1

Brixner, Dr. B., VIII-2, 3:8

Broad, W.E., I-6

Broadwell, R.P., I-6

Brock, Lt. L.V., I-14

Brockman, Henry, VIII-2

Brode, R.B., VIII-2, 3:2, 3:9

Brooks, Edwin, VIII-1

Brooks, Howard, VIII-3:3

Brooks, Lt. J.E., VIII-1

Brooks, M.L., VIII-2

Browder, Edward C., I-10

Brower, W.M., VIII-2

Brown, Capt. E.B., I-14

Brown, Edward J., I-5, IV-5, V-5

Brown, H.J., I-6

Brown, Jas. M., I-11

Brown, R.E., IV-4

Brown, Col. Staunton L., I-14

Brown, T.W., I-6, IV-5, 6

Brownell, R.M., I-4:6

Brues, Dr. A.M., I-4:6, 7

Brundage, Lt. B.M., I-4:6, 4:9, 7

Bryan, Capt. F.A., I-7

Bryant, H., II-4

Bubb, G.E., IV-5

Buchanan, J., III

Buckley, O.E., I-1

Buettner, Capt. R.J., VIII-3:8

Bugbee, S.J., IV-6

Bullard, Lt. Hudson, USN, VIII-3:5

Bullock, Lt. Col. E.F., I-5

Bullock, Maj. J.C., I-14

Bundy, Harvey H., I-4:13

Bunker, Capt. C.H., VII-1

Burbage, J.J., VIII-3:4

Burford, W.B., V-1

Burggrave, W.F., VIII-3:4

Burke, Lt. C.L., I-14

Burke, J.E., VIII-2

Burlingame, W.E., VII-2

Burman, Capt. L.C., VII-1

Burnham, Lt. B.P., I-14

Burns, J.A., IV-5

Burns, R.M., II-2

Burris, Stanley, VIII-2

Burriss, Lt. Cmdr. S., VIII-3:8

Burrough, Capt. S.E., USN, VIII-3:2

Burton, Capt. H.D., VIII-3:2

Burton, John R., Jr., I-4:7

Burton, Dr. Milton, I-4:7

Burton, R.E., IV-5

Burwell, B., VII-1,2

Bush, Lt. H.C., VIII-1,2

Bush,Lt.H.H.,I-14
Bush,Dr.V.,I-1,4:1,4:5,4:7,4:12,
    4:14,7,10,12,13,14,IV-1,2I,3,5,
    V-2,3,5,6,VII-1,VIII-2,3:9
Butler,H.G.,IV-4
Butler,S.A.,VIII-2
Byrnes,Hon.Jas.F.,I-4:7
Butler,Col.R.G.,Jr.,VIII-3:9

Cabaza,Lt.J.E.,I-14
Caldes,Wm.,VIII-2
Caldwell,I-6
Caleca,T/Sgt.V.,VIII-2
Calleghan,Lt.F.P.,I-14
Calvert,Maj.H.K.,I-14
Calvin,Capt.E.B.,I-9,12
Camac,M.,VIII-2
Cameron,Lt.C.L.,I-14
Campbell,A.W.,VIII-2
Campbell,Maj.Geo.B.,I-13
Campbell,Jos.,I-4:2
Campbell,Sir Ronald I.,I-1
Campbell,Maj.W.C.,II-3
Cannon,Rep.Clarence,I-4:1
Cantril,Dr.S.T.,I-7,IV-2II
Cappiello,Anthony,I-12
Carlberg,H.A.,IV-5,6
Carlson,B.,VIII-2
Carlson,T/Sgt.E.,VIII-2
Carlson,G.O.,IV-5
Carlson,R.,I-6
Carlson,R.W.,VIII-2
Carpenter,W.S.,Jr.,IV-1
Carr,Lt.J.F.,I-14
Carson,Ellis H.,I-6
Carter,R.C.,VIII-3:9
Casey,Lt.Col.S.B.,I-9
Cassidy,Lt.R.M.,VIII-1
Cast,P.F.,IV-6
Cater,Lt.W.D.,I-14
Catlett,Capt.W.A.,III
Center,Clark E.,II-5
Cernaghan,Capt.A.L.,VIII-1
Chadwick,Geo.VIII-2,3:9
Chadwick,Sir James,I-1,4:9,4:13,9,
    IV-1,2I,VIII-2/9

Challis,G.S.,VIII-2
Chambers,W.R.,V-6
Chapman,Dr.,V-6
Chapman,T.S.,VIII-3:4,3:7
Chapmann,Dr.S.,I-4:14,VI
Chappell,Lt.G.C.,VIII-2
Chase,Joseph,I-4:7
Cheeseman,E.L.,VIII-3:6
Cherwell,Lord,VIII-2
Chipman,E.E.,VII-1
Choate,Maj.C.E.,II-3
Christy,R.F.,VIII-2
Chubb,L.W.,I-1
Church,C.P.,IV-5
Church,H.N.,I-6
Church,P.E.,VIII-2
Churchill,Winston,VIII-2
Cieslicki,Marion,VIII-2
Clancy,Lt.J.F.,I-14
Clark,Capt.C.L.,I-14,VIII-1
Clark,J.C.,VIII-2
Clark,Dr.J.M.,VII-1
Clark,Dr.L.J.,I-4:12
Clark,O.S.,IV-5,6
Clarke,Lt.C.F.,Jr.,I-14,IV-6
Clarke,J.C.,I-5,I-6
Clarke,T.,VIII-3:7
Clarkson,Capt.W.R.,I-7
Clausen,R.E.,VIII-2
Clawson,Dr.Don,I-7
Clay,Dr.E.,I-7
Clay,Richard,VIII-3:9
Clayton,Lt.S.S.,I-14
Cleary,Pfc.Patrick,VIII-2
Cline,C.,VIII-2
Cline,Maj.V.P.,V-6
Cloud,W.T.,IV-6
Clusius,VI
Cockcroft,Prof.J.D.,I-4:9
Coe,Dr.J.R.,I-4:3
Coffin,C.S.,IV-5
Cohen,G.,VIII-3:7
Cohen,Dr.Karl P.,I-4:14,II-2
Cohn,Dr.Waldo E.,I-4:3,4:7
Colby,Dr.W.F.,I-14
Coldwell,E.S.,II-5
Cole,Dr.K.S.,I-7

Cole, Col. R. E., VIII-1
Coleman, T. J., VII-1
Collbohm, Frank, VIII-3:9
Collins, T/4 A., VIII-2
Collins, Lt. B. W., I-14
Collins, Lt. Donald C., I-4:6, VIII-3:8
Collins, Loueasy, I-10
Colton, L. J., Jr. VII-1
Compton, Dr. A. H., I-4:2, 4:4, 4:5, 4:7,
    4:9, 4:11, 4:12, 4:13, 4:14, II-3, IV-1,
    2I, 2II, VIII-3:7
Compton, Dr. K. T., I-4:2, VIII-3:8
Conant, Dr. J. B., I-1, 4:7, 4:12, 4:13,
    4:14, 7, 10, 12, IV-1, 2I, 3, 5, V-5,
    6, VII-1, VIII-1, 2, 3:2, 3:4, I-14 S2
Condon, Dr. E. U., I-4:12, VIII-2, 3:8
Cone, W. A., IV-2II
Conklin, Dr. F. R., II-5, V-1, 6
Conn, A. H., VIII-3:7
Connell, A. J., VIII-1
Connell, R. P., VIII-3:1
Conners, Lt. H. P., I-14
Connolly, Frank, I-9
Consodine, Col. Wm. A., I-14
Cook, Capt. Robt. R., I-14
Cook, Lt. Col. R. W., II-1, 5
Cook, Dr. Walter W., VIII-1, 2
Cooke, R. T., IV-6
Cooksey, Donald, I-4:13
Cooley, Lt. L. E., I-14
Cooley, Capt. T. R., USN, I-8
Coolidge, Dr. W. D., I-1
Cooney, Col. J. P., I-4:6
Cooper, Dr. C. M., I-4:4
Cooper, Capt. Murray M., I-14
Cooper, Dr. Raymond, I-7
Copps, Capt. Jas. F., I-14
Corak, Wm., VIII-3:3
Cornelius, Lt. Col. W. P., II-1, 4
Cornog, R., VIII-2
Corson, Dr. Dale R., VIII-2, 3:6
Corson, E. M., II-5
Coryell, Dr. C. D., I-4:3, 4:5
Costello, Dr. Martin, I-7
Coulson, Lt. Fred, IV-6
Coulter, L. V., VIII-3:4

Cowan, Maj. Jas. B., I-14
Cowdry, Dr. E. V., I-4:3, IV-2II
Cox, L. E. VIII-3:3
Crane, H. R., VIII-2, 3:2, 3:9
Crane, P. W., IV-6
Crawford, A. L., II-4
Crawford, Capt. R. L., I-12, II-3
Crawford, Lt. W. R., I-14
Creech, Francis L., I-6
Creighton, Lt. Esther, VIII-1
Creighton, E. J., II-2
Crenshaw, Col. Thos. T., I-12, VII-1
Creutz, Dr. E. C., I-4:4, 4:9, VIII-2
Crist, Dr. R. H., II-2, VIII-3:7, 3:9
Critchett, J. H., I-4:7
Critchfield, C. L., VIII-2
Cromer, S. J., VIII-2
Crowell, T., VIII-3:7
Crowley, Lt. J. T., I-14
Cummin, A., VIII-3:7
Cunningham, John W., IV-4
Curie, I., IV-1
Curie, M., IV-1
Curie, P., IV-1
Curran, Peter A., VIII-1
Currie, Dr. L. M., II-2
Curtis, Dr. H. J., I-7
Curtis, M. L., VIII-3:4
Curtiss, L. F., I-4:3, 4:12, VI
Custer, Gordon, III
Cydell, M. R., IV-6

Daghlian, H. K., Jr., VIII-2, 3:9
Dailey, Lt. D. H., I-14
D'Alfonso, Lt. D. V., I-14
Dallenbach, Dr., I-14
Dalton, Capt. W. E., VII-1
Dana, L. I., II-5
Daniels, Dr. Farrington, I-4:2, 4:4, IV-2I
Daniels, Lt. G. B., I-14
Daniels, H. T., IV-2II, 5
Danne, Adrian, I-4:11
Darling, L. A., IV-6
Datz, S., VIII-3:7

Davalos,Capt.,S.P.,VIII-1,2
Davidson,John E.,II-4,IV-2II,VI
Davies,Capt.John L.,I-14,VII-1
Davies,Dr.T.Harrison,I-4:7
Davis,Hon.Clifford,I-10
Davis,Eileen,VIII-3:3
Davis,Capt.Geo.B.,Jr.,I-14
Davis,John Parkes,I-4:7
Davis,Capt.N.E.,VIII-1
Davis,R.R.,VIII-2
Davis,Wm.H.,I-8
Davison,B.,VIII-2
Dawson,T/Sgt.R.,VIII-2
Dawson,Sarah,VIII-1
Day,Lt.C.E.,VIII-1
Dazzo,Lt.N.C.,I-14
Dean,Lt.Virgil F.,VIII-1
de Benedetti,S.,VIII-3:4
DeCoursey,Col.Elbert,I-4:6
DeFranco,Lt.F.J.,I-14
DeHoffman,Frederick,VIII-2
Deibner,Dr.Kurt,I-14
Deily,Lt.R.E.,I-14
Del Genio,Lt.Nicholas,I-14
DeLorimier,Col.A.A.,VIII-3:8
DeMille,J.N.,IV-6
Deming,J.C.,II-5
Dempster,Dr.A.J.,I-4:13,II-2
Demson,E.J.,VIII-2
Denfield,Rear Adm.L.E.,I-8
Denit,Gen.C.B.,I-4:6
Dennes,W.R.,VIII-2
Deringer,Dr.M.,I-7
Derry,Lt.Col.J.A.,I-9,3:8
de Seversky,Maj.A.P.,I-4:6
DeSilva,Col.Peer,I-4:6,14,VIII-2
Dessauer,Dr.Gerhard,VIII-3:8
Deutsch,Z.G.,I-4:14
Deulin,Lt.W.J.,I-14
Dewey,Bradley,VIII-3:8
DeWitt,General J.L.,IV-1
Dice,C.O.,VIII-3:4
Dick,C.H.,VIII-3:9
Dickel,G.,VI
Dickinson,Warren,VIII-3:9
Dickson,J.L.,IV-6
Dike,S.H.,VIII-2,3:9

Dill,Fld.Mar.Sir John,I-1
Dillon,J.H.,VIII-3:4
Dinius,Lt.E.F.,I-14
Dirac,P.A.M.,VIII-2
Doan,Dr.R.L.,IV-2I,2II
Dodge,B.F.,VI
Dodge,Lt.Robt.T.,I-4:7
Dodson,R.W.,VIII-2
Dole,Dr.M.,III
D'Olier,Franklin,I-4:6
Doll,Dr.E.B.,VIII-3:8
Doll,E.J.,VIII-2
Donaldson,Dr.Lauren,I-7,VIII-3:8
Donnell,Maj.A.P.,I-4:5,4:7,VIII-3:7,I-14:9
Dopel,Prof.Robt.,I-14
Dorland,Col.G.M.,VIII-3:6
Doubleday,Col.,I-4:6
Douglas,P.M.,VIII-3:1
Douglas,Dr.T.B.,VIII-3:3
Dounce,Dr.Alexander,I-7
Dow,David,VIII-2
Dowdy,Dr.A.H.,I-4:3,7
Drager,H.W.,VIII-2
Drew,T.B.,IV-6
Driscoll,M/Sgt.F.J.,I-4:6
Dubridge,Dr.L.A.,I-4:2,4:3
Dudley,Capt.L.A.,I-9
Dudley,Lt.Col.W.H.,VIII-1
Duffey,Capt.D.,VII-1
Duffield,R.B.,VIII-2
Duffiureux,Prof.,I-14
Dunbar,P.B.,I-4:3
Dunham,Lt.Robt.S.,I-13
Dunlap,R.H.,VIII-2
Dunn,Lt.J.H.,I-14
Dunning,Dr.J.R.,I-4:12,II-2,IV-1
Dunning,Maj.R.M.,I-9
Durand,Eric,VIII-3:8
DuVigneaud,Vincent,I-4:3
Dyhre,A.M.,VIII-2,3:1

Ebbs,Maj.R.F.,IV-5
Eberstadt,Ferdinand,I-4:7
Eberts,R.M.,VIII-3:7
Ebling,I-14

Foley,Melvin,VIII-2
Foltz,Lt.W.D.,I-14
Forbes,E.C.,VII-1
Ford,J.G.,V-5
Forkner,Lt.T.F.,I-14
Forney,Col.G.J.,II-5,6,V-1
Forrest,Capt.C.U.,VIII-1
Forsythe,Harry,IV-4
Forsythe,W.,VIII-3:7
Fortine,T/Sgt.F.,VIII-2
Foster,Col.Harry,I-5
Foster,Dr.Richard,I-4:1,7
Fowler,G.A.,VIII-2,3:8
Fowler,J.A.,VIII-2
Fowler,Dr.W.A.,VIII-3:2
Fox,Lt.Col.M.C.,V-6,VI
Francis,C.M.,VIII-2
Franck,Dr.J.,I-4:4,4:7
Frank,Lt.E.D.,I-7
Frankel,A.,VII-1
Frankel,S.P.,VIII-2
Fraser,Lt.Col.,H.R.,II-5
French,A.P.,VIII-2
French,Carl,VII-1
Friedell,Lt.Col.H.L.,I-4:3,4:6,
   7,VIII-3:4
Friedlander,G.,VIII-2
Friley,Dr.C.E.,I-4:11
Frisch,O.R.,VIII-2
Fritz,H.W.,III
Frolich,Lt.Col.A.J.,VIII-3:6
Froman,Dr.Darol,II-5,VIII-2
Froman,Mrs.Darol,VIII-1
Frost,H.E.,VIII-3:4
Fry,Lt.B.M.,I-14
Fuchs,K.,II-3,VIII-2
Fugard,Capt.J.R.,II-5
Fulbright,H.W.,VIII-2
Fuller,Norman G.,IV-4,6
Fulmer,Dr.E.I.,I-4:11
Fulton,Hugh,I-4:12
Funkhouser,Prof.W.D.,I-4:10
Furer,Rear Adm.J.A.,I-14,VIII-3:2
Furman,Dr.N.H.,V-2,VII-1
Furman,Maj.R.R.,I-4:6,14
Furney,Capt.Russell H.,I-14
Fussell,L.,Jr.,VIII-2

Gahan,Lt.P.J.,I-14
Gallaher,R.M.,I-10
Galloway,G.,VIII-2
Gans,D.M.,II-2
Gardner,Loris,VIII-2
Gardner,Trevor,VIII-3:2
Garner,C.S.,VIII-2
Gary,T.C.,II-3,III,IV-2II,3,6
Gaskill,Dean H.S.,I-4:11,VII-1
Geary,Lt.R.L.,I-14
Gee,Col.H.G.,VIII-1,2,3:8
Geery,Maj.B.B.,VIII-1
Generoux,R.P.,IV-5
Gentes,G.,II-5
Gentner,Dr.Wolfgang,I-14
George,A.,IV-6
George,D'Arcy,VII-2
George,Col.Warren,I-12,II-4
Gerlach,Prof.Walther,I-14
Gerow,Lt.C.E.,VI
Gessler,A.E.,II-2
Gherardi,Dr.B.,I-1
Giauque,W.F.,VIII-2
Giffels,R.F.,VIII-3:4
Giles,Lt.Gen.B.M.,VIII-3:8
Gillette,Maj.F.M.,IV-5
Gillette,Maj.Kirby M.,I-14
Gilman,Prof.Henry,I-4:14
Ginns,Dr.D.W.,I-4:9
Giordani,Dr.,I-14
Glass,E.R.,VIII-3:7
Glover,Maj.Thos.,V-6
Goering,Reichsmarshall Herman,I-14
Goldsmith,A.M.,I-9
Goldstein,E.,II-2
Goldstein,L.,VIII-2
Gomez,Elfego,VIII-1
Gonzales,Hon.Albert T.,VIII-1
Goodman,Dr.Clark,VII-2
Goodman,T/3 W.,VIII-2
Goodpasture,Dean Ernest,I-4:10
Gordon,Dr.Lincoln,I-4:7
Gore,Col.,VIII-3:9
Goring,Lt.G.E.,I-4:6
Gornall,Maj.R.F.,IV-5
Goshorn,Paul,I-5
Goudsmit,Dr.S.A.,I-14

Grace, Richard, IV-4
Grady, J.A., IV-6
Graef, R.W., VIII-1
Grafton, Capt.J.F., IV-2I
Graham, Dr.F.P., I-4:10
Graham, T., II-1, 2
Grantham, Everett M., VIII-1
Graves, A., VIII-2
Gray, C.F., I-4:11
Gray, C.W., II-5
Gray, Harvey, I-10
Greager, Maj.O.H., IV-2II, 6, VII-1
Green, C.R., VIII-2
Green, H.F., V-4
Greenberg, E., VIII-3:7
Greene, Joel, I-4:6
Greene, Priscilla, VIII-1
Greening, H.G., VIII-3:8
Grenstein, Maj.Harold, VIII-3:7
Greenwalt, C.H., II-3, IV-6, VIII-2
Greenwood, A.T., I-4:6
Grefe, F.W., VIII-1
Gregg, Lt.W.D., I-14
Gregory, Lt.Col.R.C., II-3
Greifer, Aaron, VIII-3:3
Greisen, K., VIII-2
Greninger, I-4:9
Greninger, A.B., IV-6
Grier, Herbert E., VIII-3:9
Grilli, E.R., VIII-3:3
Grizzell, Capt.R.A., I-14
Gronemeyer, F.G., VIII-3:4
Grooms, Lt.J.A., I-14
Gross, C.H., IV-5, 6
Gross, Lt.Milton, USN, I-13
Gross, Prof.Paul M., I-4:10
Groth, Dr.W., I-14
Grotjen, Capt.L.L., II-2
Groves, Maj.Gen.L.R., I-1, 4:1, 4:3, 4:4,
   4:5, 4:6, 4:7, 4:9, 4:12, 4:13, 4:14, 5,
   7, 8, 9, 10, 12, 13, 14, II-1, 3, 4, 5, III,
   IV-1, 2I, 2II, 4, 5, 6, V-1, 2, 3, 5, 6, VI,
   VII-2, VIII-1, 2, 3:1, 3:4, 3:5, 3:7, 3:8,
                                      3:9,
                              I-14 S2, S3
Guarin, Maj.Paul L., VII-2
Guilfoyle, Lt.T.D., I-14
Gunn, Dr.R., I-4:12, 4:14
Gunnes, R.C., VIII-3:7

Gunter, Thurman, VIII-1
Gufinski, D.H., VIII-2
Gurley, M.H., VIII-1
Gustavson, Dr.R.A., I-4:2
Gysae, Dr., I-14

Haeg, R., VIII-3:3
Hadlock, Maj.Canfield, I-14, V-6, VII-1
Hageman, Capt.P.O., I-4:6, VIII-1, 2
Hageman, R.C., IV-5, 6
Haggerson, F.H., II-5
Hahn, Prof.Otto, I-14, II-1, 2, IV-1, 2I
Haley, Capt.Jas.W., I-14
Halban, Dr.H., I-14, 4:9
Halifax, Earl, I-1
Hall, David, VIII-2
Hall, Herbert, VIII-3:8
Hall, Capt.Howard F., I-14
Ham, Maj.R.G., I-14
Hamilton, Jos.G., I-4:3, 7, VIII-3:8
Hammel, E.F., VIII-2
Hammett, Dr.L.P., VIII-3:9
Hamming, R.W., VIII-2
Humphries, Thuron, I-4:6
Hancock, J.M., I-4:7
Handley, R.W., VII-2
Hane, Wilbur, VIII-2
Hanig, M., VIII-3:7
Hanngock, J., II-5
Hanson, B.F., IV-6
Hanson, Capt.E.L., USN, VIII-3:2
Hare, R., IV-6
Hargrave, V-6
Haring, Dr.M.M., VIII-3:4
Harkings, Dr.W.D., I-4:14, II-2
Harlin, R.H., I-6
Harman, Col.J.M., I-5, VIII-1, 2
Harmon, James, VIII-1
Harmon, K.M., VIII-2
Harmon, CWO R.C., VIII-1
Harms, T/3 D., VIII-2
Harned, Prof.H.S., I-4:14, VII-1
Harris, L.J., IV-5
Harris, R.I., IV-6
Harrison, Geo.L., I-4:13

Harrison,Dr.G.R.,VII-1
Harrison,Dr.W.N.,I-4:12
Harshaw,W.J.,VII-1
Hart,Maj.Gen.G.T.,VIII-1
Hart,Lt.H.S.,I-14
Harteck,Prof.P.,I-14
Hartman,W.C.,II-5
Hartridge,A.L.,III
Hartwig,Dr.,I-14
Hastings,A.B.,I-4:3
Hatch,Senator Carl,VIII-1,3:8
Hauelsen,B.R.,VIII-3:9
Hauth,W.E.,II-5
Haupt,L.,IV-5
Haven,Dr.Francis,I-7
Hawkins,D.,VIII-2
Hawkins,L.G.,VIII-2
Hayden,Lt.A.S.,I-9
Hayes,Capt.Arthur J.,I-14
Hayward,Capt.J.T.,USN,VIII-3:2
Hearon,Capt.W.M.,VII-1
Hecker,Dr.J.C.,V-6
Hecker,Capt.M.L.,VII-1
Heflin,Col.,VIII-3:9
Heilpern,Maj.Bert H.,I-5
Heisenberg,Prof.Werner,I-14
Helmholtz,L.,VIII-2
Hempelmann,Dr.L.H.,VIII-1,2
Henderson,R.W.,VIII-2,3:8
Henne,A.L.,II-2
Henning,C.O.,IV-5
Henningsen,E.V.,IV-6
Hensen,J.R.,I-6
Henshaw,Dr.P.S.,I-4:6
Hensley,J.W.,I-11
Herb,R.G.,VIII-2
Hersey,John,I-4:6
Hertel,Prof.K.L.,I-4:10
Hertz,G.,II-2
Heston,Dr.W.E.,I-7
Hewett,Brig.Gen.Hobart,VIII-3:8
Heyd,J.W.,VIII-3:4
Heydenberg,N.P.,VIII-2
Hiby,Dr.,I-14
Higginbotham,W.A.,VIII-2
Hightower,L.E.,VIII-2
Hilberry,Dr.Norman,I-4:2,4:13,IV-1,
  21,6

Hill,Dr.A.,VII-1
Hill,C.F.,V-5
Hill,Nathaniel,I-4:6
Hill,Capt.R.C.,II-3,V-4,VIII-2
Hillman,G.E.,IV-5
Hinch,Wm.H.,VIII-2
Hirschfelder,J.O.,VIII-2,3:8,3:9
Hitchcock,Lt.G.E.,I-14
Hittell,J.L.,VIII-2
Hochwalt,Dr.C.A.,VIII-3:4
Hodge,Dr.Harold,I-7
Hodges,Lt.G.H.,I-14
Hodgson,Col.John S.,I-12
Hoecker,Dr.,I-14
Hoffman,Dr.J.I.,I-4:12,VIII-2
Hogness,Dr.T.R.,I-4:4,4:9,4:13,14,
  IV-21,
Holder,Lt.Thornton F.,USN,I-13
Holifield,Rep.Chet,VIII-3:8
Holladay,J.A.,VII-1
Holland,Lt.H.H.,I-14
Holloway,M.G.,VIII-2,3:8,3:9
Holmes,A.R.,VII-1
Holmquist,V.E.,IV-6
Holt,Dr.Harold,I-7
Holt,Dr.W.B.,I-7
Holton,Dr.W.B.,I-4:12,13
Hood,C.B.,VIII-3:3
Hoogterp,J.J.,VIII-2
Hookway,R.E.,VIII-3:3
Hoover,Cmdr.G.O.,I-1
Hoover,Vice Adm.J.H.,VIII-3:8
Hopkins,H.S.,III
Hopper,Lt.J.D.,VIII-2
Horan,Capt.John A.,I-14
Horine,Rosella,VIII-3:3
Hornberger,Carl,I-4:6
Horowitz,Lt.J.J.,VIII-1
Hossfeld,Lt.Cmdr.Raymond F.,I-13
Hostetter,G.M.,I-11
Hough,Maj.Benjamin,I-12,II-2,V-3
Houghton,B.,VIII-2
Houston,G.C.,IV-6
Houston,Lt.L.O.,I-14
Houtermans,Prof.,I-14
Hovde,Dr.F.L.,VIII-3:2
Howard,John B.,I-4:7
Howard,Dr.L.H.,V-4

Howe,C.D.,I-1
Howe,J.P.,IV-6
Howland,Lt.J.W.,I-4:6,7
Howson,Lt.Chas.,I-13
Hoyt,F.C.,II-5,VIII-2
Hoyt,Henry,VIII-2
Hubbard,D.O.,II-5
Hubbard,J.M.,VIII-2
Huber,A.P.,II-5
Huene,Lt.T.F.,VIII-1
Hueper,Maj.Gen.H.P.,I-5
Huffines,R.A.,VIII-2
Huffman,Dr.J.R.,I-4:9
Huffman,J.W.,I-4:4
Hughes,A.L.,VIII-2
Hughes,Jas.,VIII-2
Hughie,Lt.E.V.,VIII-1
Huisking,Lt.W.W.,I-14
Hull,Maj.A.C.,Jr.,VIII-2
Hull,Dr.D.C.,V-6
Hull,D.E.,II-5
Hull,Dr.H.L.,V-6
Hull,Lt.J.M.,I-14
Humes,W.B.,II-5
Hussey,Rear Adm.G.F.,VIII-3:2,3:8
Hutchins,Chancellor Robt.M.,I-4:2
Hutchinson,Lt.Col.W.S.,I-4:5,4:7
Hutchison,C.A.,VIII-3:7

Ihwe,Dr.,I-14
Imrie,Maj.Mathew,VIII-1
Ingham,Sidney,VIII-2
Inghram,M.G.,VIII-3:7,3:9
Ingliss,D.R.,VIII-2
Ingraham,Henry G.,I-4:7
Ireland,Capt.J.D.,V-2
Irvine,Dr.John W.,I-4:7
Irwing,W.,IV-2II
Iwase,Dr.,I-14

Jackson,Lt.Col.,Geo.A.,I-6
Jackson,Maj.Joseph J.,I-14
Jackson,LeRoy,I-12

Jackson,Lt.R.G.,I-14
Jacobs,S.J.,VIII-3:9
Jacobson,Dr.Leon,I-7
Jacobson,Dr.Louis,X-7
Jacobson,Dr.L.O.,I-4:9,4:13
Jaffe,Capt.L.,I-7
Jameson,Capt.Lloyd,I-7
Jansen,Dr.,I-14
Jefferies,Zay,I-4:3,4:7
Jenike,Capt.W.F.,VIII-1
Jenkins,H.P.,V-4
Jennings,Hon.John,Jr.,I-10
Jepson,Lt.Morris,VIII-2
Jerrems,A.S.,VIII-3:6
Jett,Lt.S.K.,I-14
Jette,E.R.,VIII-2,3:4
Jewett,Dr.F.B.,I-4:3,I-1
Johannesson,Maj.R.E.,I-14,II-5
Johns,Dr.I.B.,I-4:11,VIII-2
Johnson,Lt.Col.A.C.,I-9,V-4,VIII-3:2,
  3:3
Johnson,Albin E.,I-4:7
Johnson,A.R.,VIII-2
Johnson,Lt.C.L.,I-14
Johnson,C.M.,I-12
Johnson,E.A.,II-3
Johnson,Lt.Henry,USN,I-13
Johnson,Capt.H.R.,VIII-3:6
Johnson,Dr.J.R.,I-14
Johnson,Capt.L.E.,I-14,IV-5,I-14 S3
Johnson,R.W.,VIII-3:1
Johnson,Dr.W.C.,I-4:5,IV-2II
Johnson,Lt.Wm.G.,I-11
Johnston,Prof.H.L.,VIII-2,3:3
Joliot,Prof.F.,I-14,IV-1,VIII-2
Jones,E.L.,II-4
Jones,N.H.,II-4
Jones,T.A.,III
Jones,Maj.T.O.,I-14
Jones,T.R.,VIII-3:7
Jones,Wesley M.VIII-3:3
Jorgensen,T.A.,VIII-2
Joris,G.G.,II-2
Judson,C.M.,VIII-3:7
Junkin,A.V.,II-4
Junod,Dr.,I-4:6
Justi,Dr.,I-14

Kadlec, Lt. Col. H. R., IV-5
Kaffman, Cmdr. D. L., VIII-3:8
Kahn, Milton, VIII-3:9
Kalichman, E., VIII-3:7
Kamen, Martin, VIII-3:9
Kammer, Dr. Adolph, I-7
Karl, Capt. Clarence L., I-4:7, VIII-3:1
Kasner, Cpl., I-4:6
Kaspar, Col. E. J., VIII-3:4
Kay, W. C., IV-2II, 6
Kearton, C. F., II-3
Keeler, Leonarde, V-6
Kehl, C. L., VIII-2
Kehoe, R. A., VIII-3:4
Keiser, Maj. Eubert, VII-2
Keith, P. C., II-5
Keller, John, VIII-2
Keller, Jos. M., I-4:11, VIII-2
Kelley, Armand, VIII-2
Kelley, Dr. R. E., VIII-3:4
Kelley, Lt. R. H., I-14
Kelley, Lt. Col. Wilbur E., I-4:7, 12,
    V-1, 2, 3, 4, 6, VII-1
Kellogg, J. M. B., VIII-2
Kelly, D. N., I-11
Kelly, Jos. A., VII-1
Kendall, Dr., I-4:14
Kendall, Lt. R. S., I-14
Kennedy, Dr. Jos. W., I-4:3, IV-1, 2I,
    VIII-2, 3:3, 3:4
Kent, R., 3d, VII-1
Kershaw, Stanley, VIII-1, 2
Kerst, D. W., VIII-2
Kibnick, A., VIII-3:7
Kilgore, Sen. Harley M., I-4:1
Killough, Lt. R. S., I-14
Kilpatrick, Dr. M., VIII-3:7
King, Admiral E. J., IV-5, VI, VIII-2, 3:8
King, Dr. E. Q., I-4:3
King, Capt. J. A., I-14
King, Capt. J. E., III
King, L. D. P., VIII-2, 3:9
King, Lt. W. A., I-14
King, Prime Minister W. L. McKenzie,
    I-4:7
Kingdon, Dr. K. H., I-4:14, V-2
Kinsey, H. D., II-5

Kinzel, Dr. A. B., I-4:7
Kipp, I-6
Kirchner, Prof. F. F., I-14
Kirkman, Capt. R. W., I-4:7, 14, VIII-3:2
Kirkpatrick, Col. E. E., I-4:3, II-5,
    VIII-2, 3:4
Kirshenbaum, A. D., VIII-3:7
Kirshenbaum, Dr. I., VIII-3:7
Kistiakowsky, G. B., VIII-2, I-1, VIII-3:9
Klaproth, M. H., VII-1
Klarquist, Kenneth, I-13
Klav, Abel, I-6
Klein, A. C., V-1, 3
Klemm, Dr., I-14
Kline, S. A., VIII-2
Klossner, Maj. R. H., II-5
Knowles, Col. Miles, I-4:1
Knudsen, H. T., I-6
Koenig, C. J., I-6
Koester, G. A., VIII-3:8
Kohl, Dr., I-14
Konopinski, E. J., VIII-2
Kopfermann, Prof., I-14
Kopp, M., VIII-3:7
Koranda, Lt. Hugo, I-14
Korshing, Dr., I-14
Koski, W., VIII-2
Kowerski, Dr. Lew, I-14
Kramer, A. G., II-5
Kraus, Prof., I-4:14
Krevatch, W. B., IV-5
Kron, Dr. G. E., VIII-3:2
Krous, C. A., V-1
Kruger, W. C., VIII-1, 2
Kuhn, Prof. Richard, I-14
Kupferberg, T/Sgt. C., VIII-2
Kurpka, M., VIII-3:7
Kurtz, A., VIII-3:7
Kyle, Col. W. E., I-4:13

LaBine, Gilbert A., VII-1
LaCroix, Col. James P., I-5
Lail, G. G., IV-1, 6
Lamie, R. S., VI
Lanahan, T. B., VIII-3:8

Landrum, C.U., IV-4
Landshoff, R., VIII-2
Lane, Cmdr. Donald E., I-13
Lane, Dr. J.A., I-14
Langan, E.F., I-9
Langer, L., VIII-2
Langham, Dr. Wright, VIII-2
Lannon, Lt. J.J., I-14
Lansdale, Col. John, Jr., I-14
Large, Capt., VIII-2
Largey, J.R., II-5
Lark-Horovitz, VIII-3:9
Larkin, Capt. R.A., VIII-2
Larkin, T/Sgt. W., VIII-2
Larsen, E.M., VIII-3:4
Larson, Dr. C.E., V-6
Lauder, D.H., IV-1,6
Lauritsen, Dr. C.C., VIII-2, 3:2
Lauritsen, Dr. T., VIII-3:2
Lautenslager, Murice, VIII-3:3
Lavender, Capt. Robt. A., USN, I-13
Lawrence, Dr. E.O., I-4:5, 4:7, 4:12, 4:13,
    IV-1, V-1, 2, 3
Lawrence, W.L., I-6
Lawson, Philip, VIII-2
Leahy, Maj. P.C., VII-1
Leahy, Lt. T.H., I-14
Leahy, Adm. W.D., VIII-3:8
Lee, C.E., VIII-3:1
Lee, VO. Maynard E., I-14
Leet, L.D., VIII-2
Lefler, B.R., VIII-1
Leggo, Dr. Christopher, I-7
Lehr, H.K., VIII-3:8
Lehy, Lt. Robt. J., USN, I-13, VIII-2
Leitz, F.J., Jr., VIII-3:4
LeMay, Maj. Gen. C.E., VIII-2, 3:8, 3:9
Lenzie (Flanagan, Prior and), IV-4
Lemons, J.F., VIII-2
Leonard, Capt. Geo. B., I-14
Leonard, Robt., I-4:6
Lesco, Dr. Herman, I-7
Leverett, M.C., VIII-3:4, 4:4
Levin, A.A., VII-1
Levin, S., VIII-3:7
Levine, Phil, I-4:6
Lewis, Dr. W.K., I-1, 4:2, 4:4, II-3, III,
    VIII-2

Libby, Dr. W.F., I-4:5, II-2
Liberatore, L.C., VIII-3:7
Lilienthal, David, I-4:2, 4:7
Linares, A., IV-6
Lincoln, Brig. Gen. G.A., VIII-3:8
Lindemann, Dr. F.A., I-4:14
Lindsay, Franklin A., I-4:7
Lindvall, Dr. F.C., VIII-3:2
Linenberger, Dr. G.A., VIII-3:8
Linschitz, H., VIII-2
Lipkin, David, VIII-2
Lister, Gordon K., I-13
Little, T/3 F.L., VIII-2
Little, Norman M., IV-4
Littler, D.J., VIII-2
Livingston, Capt. Claude, I-9
Livingston, D.C., VIII-2
Livingston, Prof. M.S., I-4:2
Llewellin, Col. J.J., I-1
Lockhart, C.C., IV-6
Lockridge, Lt. Col. E.W., VIII-2, 3:1,
    3:6, 3:9
Lofgen, Dr., V-3
Lokietz, M.S., VII-1
Long, E.A., VIII-2
Long, K.E., VII-1
Longsworth, Dr. L.G., I-4:14
Loomis, Dr. F.W., I-4:2
Loomis, R.L., VIII-2
Lord, Capt. W.W., I-12, V-3
Lorenz, Dr. E., I-7
Lovejoy, V-6
Lowe, Brig. Gen. Frank, I-4:1
Lubeck, W.A., I-6
Luke, Maj. C.D., II-5
Lum, Dr. J.H., I-4:4, IV-2II, VIII-3:4
Lynch, E.R., V-4
Lyon, Capt. C.M., USN, VIII-3:8
Lyster, T.L.D., II-5

MacCready, W.H., IV-6
MacDougall, Dr.D.P., VIII-3:9
Machen, A., VIII-2, 3:6, 3:8
MacInnes, Dr.Duncan, I-4:14
Mack, E., Jr., II-2
Mack, J.E., VIII-2
MacKenzie, Dean C.J., I-1, 4:9
Mackey, B.H., IV-6
Mackie, F.H., IV-2II, 5
MacLeod, W.J., I-6
MacWood, Dr., VIII-3:3
Maddy, J.R., I-11
Madigan, Lt.A.E., I-14
Madorsky, Dr.S.L., I-4:14
Magee, J.L., VIII-2, 3:8
Maguire, Lt.J.J., I-14
Mahon, Rep.Geo.E., I-4:1
Mahoney, Lt.J.H., I-14
Maider, J.E., IV-6
Makins, Roger, I-4:13
Manley, Dr.John H., I-4:5, VIII-2, 3:9
Mansfield, Col.H.W., VIII-3:8
Manvel, F.I., V-5
Maples, Chief H.H., I-12
Marden, Dr.J.W., VII-1
Mark, Carson, VIII-2
Marley, W.G., VIII-2
Marsden, Col.E.H., I-5, 6, IV-6
Marsden, N.H., I-11
Marsh, Lt.Harold, I-14
Marshak, R., VIII-2
Marshall, D.G., VIII-2
Marshall, Gen.Geo.C., I-1, 4:1, IV-1,
    VIII-3:8, I-14 S2
Marshall, Col.J.C., I-1, 5, II-5, III,
    IV-1, 2I, V-2, 4, 6, VII-1
Marshall, S., VIII-2
Martel, J.P., IV-2II
Martin, D.S., VIII-2
Martin, Rep.J.W., Jr., I-4:1
Mastick, Ens.D., VIII-2, 3:8
Mateer, W.D., VII-2
Mathe, Maj.Robt.E., I-14
Mather, Jas, VIII-3:7
Mattauch, Dr., I-14
Matthews, Lt.Col.Clyde, I-4:6

Matthews, R., VIII-2
Matthias, Col.F.T., I-4:1, IV-1, 5
Mattox, Sgt.James, VIII-1
Maurer, Dr.Werner, I-14
May, A.I., VIII-3:9
May, Lt.J.P., I-14
McAdam, Lt.R.G., I-14
McBee, E.T., V-1
McCarthy, E.C., VIII-3:4
McCaskill, Lt.E.A., I-14
McCaulay, D.A., VIII-3:7
McCauley, Capt.W.R., I-5
McClenahan, Capt.Henry I., I-4:6, 14
McCloy, John J., I-4:7
McComb, Wm.R., I-8
McCord, W.O., VIII-3:8
McCormack, James K., I-6
McCormack, Rep.John W., I-4:1
McCormick, Capt.J.D., II-3
McCormick, Maj.J.L., Jr., II-3
McCullough, Dr.C.R., I-4:4, 4:7
McCune, S.W., Jr., VII-1
McDaniel, Dr.P.W., I-4:10
McDermott, Lt.E.P., USN, I-9
McDonald, Dr.Ellice, I-7
McDonald, F.H., IV-5
McGavock, Maj.J.H., VIII-1
McKee, J.F., VII-1
McKee, R.E., VIII-1, 2
McKenzie, J.G., I-10
McKenzie, Capt.R., VII-1
McKeon, Francis D., V-3, VIII-3:4
McKibben, J.L., VIII-2
McKinley, Maj.J.H., IV-2I, VII-1
McKinney, Maj.Russell L., I-14
McKneight, S.A., IV-6
McLeod, Capt.R.J., I-14
McMahon, Senator Brien, VIII-3:8
McMillan, Dr.E.M., IV-1, 2I, VIII-2, 3:3, 3:8
McMullen, Lt.Wm.E., I-14
McNally, Dr.J.G., V-6
McNierney, W.L., VIII-2
McVeigh, T.F., II-4
Mead, Sen.Jas.M., I-4:1
Means, M.G., II-5
Mears, Capt.B.J., I-7

Meder, Capt. Jos. F., I-14
Meier, G. F., IV-5
Meigs, Douglas P., I-6
Meikle, G. Stanley, VIII-3:9
Meints, Capt. Ralph, VIII-3:4
Mengerink, Maj. C. E., VIII-3:1
Menke, Capt. B. W., I-14, VII-1
Menzer, Dr., I-14
Merritt, Maj. P. L., VII-1
Metcalf, Col. Herbert E., I-13
Metcalf, Capt. Roger, I-7
Metz, Chas. F., VIII-2
Metzger, Lt. D. A., I-14
Michenor, W. F. X., I-4:7
Middleton, A. E., VIII-3:9
Miller, Dr., V-5
Miller, G. Mark, IV-4
Miller, D. A., IV-6
Miller, H. A., IV-6
Miller, H. I., VIII-2
Miller, Maj. J. F., I-10
Miller, J. W., VIII-1
Miller, Maj. R. H., I-9
Miller, Lt. j. g. V., VIII-2
Miller, W. C., I-4:6
Millett, K., IV-5
Mills, F., II-5
Mills, Harvey, I-11
Milton, W. H., Jr., IV-6
Mims, M. P., VIII-3:9
Minikes, Lt. Col. Solette E., I-14
Minissini, Vice Adm., I-14
Miskulin, Lt. Mike, I-14
Mitchell, D. P., VIII-2, 3, I-4:12
Moeding, Henry, VIII-2
Moeller, Commo. L. N., VIII-3:2
Moliere, Dr., I-14
Monier, Dr. J. A., III
Monro, D. A., VIII-3:7
Montgomery, S. L., VIII-3:7
Montoyo, Adolfo, VIII-1
Montoyo, Ernesto, VIII-1
Moon, P. B., VIII-2
Moon, W. F., VIII-2
Moore, Maj. D. C., V-6
Moore, Cmdr. Hudson, I-9, VIII-2

Moran, Maj. J. J., II-1, 3, 5
Morgan, Capt. A. N., USN, VIII-3:8
Morgan, Elmo, VIII-1
Morgan, Fred, I-10
Morgan, J., I-6
Morgan, Dr. K. Z., I-7
Morrell, Capt. V. J., IV-6
Morrison, Dr. Philip, I-4:4, 4:6, 4:7, 14, II-5, VIII-2
Morrissey, Geo. E., I-6
Morse, Dr. P. M., I-4:2
Morse, Capt. R. D., VII-1
Moses, Brig. Gen. R. G., I-1
Mosgrove, Lt. Richard, VIII-1
Moshier, R. W., VIII-3:4
Motichko, T/Sgt. L., VIII-2
Motohashi, Dr. N., I-4:6
Moulton & Powell, IV-4
Mountjoy, Capt. P. B., I-14, IV-5, 6
Mueller, Maj. Walther E., I-13
Mulliken, Dr. R. S., I-1, 4:5, 4:13, 4:14
Mullin, Dr. H. R., I-4:12
Mulvihill, Lt. Helen E., VIII-1
Muncy, J. A. D., VIII-2
Murphree, Dr. E. V., I-1, 4:12, 4:14, II-3, III
Murphy, Lt. E. F., Jr., I-14
Murphy, Maj. E. J., IV-2II
Murphy, G. M., II-2
Murphy, T/Sgt. W., VIII-2
Murray, Capt. Jas. S., I-14
Musser, J. N., IV-6
Musser, Capt. Sam., VIII-3:6
Myers, Jacob, VIII-3:3

Nash, Lt. Clinton, VIII-1
Nauman, Col. A. C., VIII-1, 2
Naylor, E. E., I-5
Neblett, Judge Cohen, VIII-1
Neddermeyer, Seth, VIII-2
Nedzel, Dr. B. A., VIII-3:8
Neel, Lt. J. V., I-4:6
Nelson, Lt. Col. C. A., I-5, 8
Nelson, Donald N., I-9
Nelson, E. C., VIII-2

Nelson, Dr. Harold, I-7
Nereson, Dr. N. N., VIII-3:8
Netcher, Col. E. G., I-5
Neuert, Dr. Hugo, I-14
Neverick, Lt. A. A., I-14
Newburger, Maj. Sidney, VIII-2
Newcomb, Maj. R. I., IV-5, VIII-3:1
Newell, E. E., I-4:9
Newell, W. S., VIII-3:8
Newlon, C. E., II-5
Newman, Brig. Gen. J. B., Jr., I-4:6,14
Newson, E. W., I-4:4
Newton, Amos S., I-4:11
Nichols, Brig. Gen. K. D., I-1, 4:2, 4:3,
    4:4, 4:6, 4:7, 4:10, 5, II-5, III,
    IV-1, 6, V-2, 4, 6, VI, VII-1, 2,
    VIII-3:4, 3:7
Nickman, A. A., II-4
Nickson, Dr. J. J., I-7
Nickson, Dr. J. M., I-7
Nickson, Dr. Margaret, I-7
Nielsen, Col. A. W., I-4:9
Nielsen, M. L., VIII-3:4
Nier, Dr. A. O. C., I-4:12, 4:14, II-2
    III, V-2, VI, VIII-3:9
Nininger, Lt. Robt. D., VII-2
Nitze, P. H., I-4:6
Noble, Lt. Col. C. C., I-5
Noel, C. K., VIII-1
Nolan, Lt. J. E., I-14
Nolan, Capt. Jas. F., I-4:6, VIII-1, 2
Nolan, T. B., I-9
Nooker, T/Sgt. E., VIII-2
Nordheim, Gertrude, VIII-2
Nordheim, Dr. Lothar W., I-4:4, 4:7,
    VIII-2
Norris, E. O., II-2
Norris, J. C., I-10
Norstad, Gen. L., VIII-3:9
Norton, Dr. B. M., VIII-3:2
Norton, F. H., VIII-2
Norwood, Dr. W. D., I-7, IV-6
Notman, D. O., IV-6
Nuessle, Gustave, VIII-3:3
Nunn, E. F., I-11

Oberbeck, Lt. Col. A. W., VII-1, 2
O'Brien, Lt. C. H., I-14
O'Brien, Cmdr. Gerald, I-13
O'Brien, Inez, VIII-2
O'Brien, Col. J. J., I-10, IV-4
O'Byrne, M. J., I-10
O'Connell, Lt. J. J., I-14
Ofstie, Rear Adm. R. A., VIII-3:8
O'Gara, Lt. J. L., I-14
Ogle, Wm., VIII-2
Ohlinger, Dr. L. A., I-4:4, 4:9
O'Keefe, Ens. B., VIII-2
Old, Lt. Cmdr. Bruce S., I-14
Olmstead, T. H., VIII-2
Olsen, E. E., VIII-2
Olson, Lt. E. G., I-14
O'Malley, Lt. Thos. R., USN, I-13
O'Meara, Capt. P. F., I-12, V-4
Oppenheimer, Frank, VIII-2
Oppenheimer, Dr. J. R., I-1, 4:5, 4:7, 4:13,
    II-5, 6, VIII-1, 2, 3:1, 3:2, 3:3, 3:7, 3:9
Orfalea, Lt. G. A., I-14
Osborne, R. B., VIII-1
Osenberg, Prof. Werner, I-14
Otto, F., IV-6
Oughterson, Col. A. W., I-4:6
Overbeck, W. P., IV-6
Overstreet, Dr. Ray, I-7

Page, Lt. Col. Gordon B., VII-2
Pagett, Lt. A. A., Jr., I-14
Palmer, Maj. T. O., VIII-2
Paolino, Lt. A., I-14
Pardee, F. W., Jr., IV-2II, 3, 6
Parker, H. M., I-7
Parkinson, T. I., Jr., I-4:2
Parratt, L. G., VIII-2
Parrish, D. T., VIII-3:4
Parsons, Col. W. B., I-14
Parsons, Rear Adm. W. S., USN, VIII-2, 3:2,
    3:5, 3:8, 3:9
Pash, Lt. Col. Boris T., I-14
Pasquier, L. P., II-5
Patterson, Robt. P., I-8, 9, IV-1, 4, 5, V-5
Paulus, M. G., VIII-3:7

Schacher, Lt. G.P., I-14
Schaeffer, Lt., I-4:6
Schaffer, Lt. W.F., VIII-2
Schiercke, Sgt. T.H., V-4
Schmeidel, J.R., IV-5
Schmitz, H.E., I-6
Schoppe, Margaret, VIII-1
Schreiber, R.E., VIII-2, 3:8, 3:9
Schrenk, Dr. H.H., I-7
Schruben, O.H., VIII-3:1
Schultz, Gus H., VIII-2
Schumann, Prof. Erich, I-14
Schumann, Lt. V.K., I-14
Schumb, Prof. W.C., VII-1
Schutz, P.W., VIII-3:7
Schwab, W.G., VIII-3:7
Schwartz, H., VIII-3:7
Schwartz, J.I., VIII-2
Schwartz, R., VIII-3:7
Schwartz, Dr. S., I-7
Schwartz, Lt. Samuel, I-14
Schwellenbach, Judge L.B., IV-4,6
Schwertfeger, A.J., IV-6
Scott, D.L., VIII-3:4
Scott, Dr. Kenneth S., I-7, VIII-3:8
Scott, Capt. Stuart W., I-13
Scoville, Dr. Herbert, Jr., VIII-3:8
Scribner, Dr. B.F., I-4:12, VII-1
Seaborg, Glenn, I-4:5, 4:13, IV-1, VIII-2
Searls, Fred, I-4:7, VIII-3:8
Seaton, Warren, I-13
Seckendorff, E.W., V-3
Seelman-Eggebert, Dr., I-14
Seely, L.B., VIII-2
Seeman, Col. L.E., VIII-1, 2, 3:6
Segre, Emilio, IV-1, VIII-2, 3:9
Seider, Capt. R.G., II-3
Seitz, Maj. B.G., I-5
Seitz, Dr. Frederick, I-4:4, 4:10, IV-2II
Selfridge, G.C., VII-1, 2
Seminara, Lt. L.A., I-14
Semple, Capt. David, VIII-2
Sentiff, H.J., VII-1
Serber, C.L., VIII-2
Serber, Dr. R., I-4:6, VIII-2
Serduke, J.T., VIII-2
Sergeant, Maj. Wm. T., I-14
Seubert, E.G., VIII-3:7

Seybolt, A.U., VIII-2
Shadel, Henry K., VIII-1
Shane, C.D., VIII-2
Shapiro, M.M., VIII-2, 3:8
Shaughnessy, John, I-6
Shaw, Col. David F., I-14
Shaw, Capt. Robt. T., I-14
Sheahan, R.D., I-9
Sheard, H., VIII-2
Sheldon, G.T.E., II-5
Shepherd, C.H., IV-5
Shields, Maj., VIII-3:2, 3:9
Shipman, I-4:1
Shodlesky, J., VIII-3:7
Short, Hon. Dewey, I-10
Siemes, Father J.A., I-4:6
Silverman, M.L.B., I-7, VIII-3:4
Simmons, S.J., VIII-2
Simon, W.O., IV-6, I-4:1
Simons, Capt. Foyle W., I-14
Simpson, C.S. IV-5
Simpson, Dr. O.C., I-4:4, 14
Simpson, Capt. Wm. J., I-14
Simpson, Dr. W.L., I-4:3
Sinsi, Dr. Nathan, I-7
Skaff, P.S., IV-6
Skinner, Lt. Col. H.E., IV-6
Sklaire, Dr. Harris, I-7
Skyrme, T.H.R., II-3, VIII-2
Slack, Prof. F.G., I-4:10, II-2
Slater, Dr. J.C., I-1
Slotin, Louis, VIII-2
Slotin, Sonia, VIII-2
Smith, Maj., VIII-3:9
Smith, Lt. A.E., I-14
Smith, C.S., VIII-2, 3:4, 3:7
Smith, E.C., VII-2
Smith, Maj. Francis, I-14
Smith, Lt. H.R., I-14
Smith, J.S., VIII-3:7
Smith, Lt. L.D., I-14, VIII-3:8
Smith, M.H., IV-6
Smith, Lt. Col. Ralph C., I-13, VIII-1, 2, 3:2
Smith, R.J., I-4:6
Smith, Wilbur F., I-13
Smyth, Dr. C.P., I-14
Smyth, Prof. H.D., I-1, 4:2, 4:5, 4:12, 4:13, 4:14

Snell, Dr. A. H., I-4:3
Snoddy, Prof. L. B., I-4:14
Snow, Lt. Edgar, USN, I-4:6
Snyder, Capt. Floyd, VIII-1
Snyder, Rep. J. Buell, I-4:1
Soddy, F., IV-1
Solberg, Rear Adm. T. A., VIII-3:8
Somervell, Gen. Brehon, I-1, IV-1
Southerland, Capt. J. E., I-14
Sowa, Frank, I-4:6
Sox, Lt. Col. C. B., I-7
Spaatz, Lt. Gen. C. A., VIII-3:8
Spalding, Leslie F., I-6
Sparkman, Hon. John, I-10
Spedding, Dr. F. H., I-4:2, 4:5, 4:7, 4:11,
    VII-1, VIII-2
Spence, R. W., VIII-2
Spencer, Dr. R. R., I-7
Sperling, C. C., II-5
Spicka, J. J., VIII-3:4
Spovack, J. S., VIII-3:7
Sprouse, John D., I-10
Spyridakis, Lt. E. G., I-9
Squires, A. N., II-5
Squires, Dr. Wm., I-7
Stagg, Maj. W. W., VII-1
Stallings, Charlie, VIII-2
Stallings, J. H., I-7
Stanford, R. E. L., IV-6
Staniforth, R. A., VIII-3:4
Stannard, Capt. W. B., I-14
Stansbury, Capt. Max E., I-14
Stapleton, Newton, I-7
Stapleton, T. M., IV-6
Starke, H. H., VIII-3:7
Starr, Dr. C., I-4:7
Staub, H. H., VIII-2
Steadman, Dr. L. T., I-7
Stellwagen, I-6
Sterling, Capt. Thos. A., I-14
Sterner, Dr. James H., I-7
Sterns, J. C., IV-21
Stetter, Dr. G. K., I-14
Stevens, Maj. W. A., VIII-1, 2
Stevenson, Lt. Cmdr. E., VIII-2
Stevenson, J. H., VIII-2
Stewart, Dr. Irwin, I-12, 13, VIII-3:2
Stewart, Lt. Col. S. L., VIII-1, 2, 3:1,
    3:2, 3:7, 3:9

Stilwell, Gen. J. W., VIII-3:8
Stimson, Sec. H. L., I-1, 4:1, 4:13, IV-1
Stockinger, Dr. H. E., I-7
Stone, Dr. R. S., I-4:7, 7
Stout, J. W., VIII-2
Stow, F., VIII-3:7
Stowers, Maj. D. M., VII-1
Stowers, Lt. Col. J. G., II-1, 3, 5,
Strassmann, Dr. F., I-14, II-2, IV-1, 21
Streeter, Dr. A. N., I-7
Streigleder, C. J., VIII-3:4
Streit, Frank, V-6
Strod, A. J., VII-1
Stroke, F., VIII-2
Strong, Maj. Gen. Geo. V., I-14
Strum, A. E., VIII-1
Sturges, Maj. D. G., VII-1
Sturges, S., VII-1
Sturgis, N. D., IV-5, 6
Sturgis, Roscoe G., I-10
Styer, Maj. Gen. W. D., I-1, III, VIII-2
Suhr, Dr., I-14
Sulerud, Lt. J. C., I-14
Sullivan, C. A., IV-6
Sullivan, S. P., VII-1
Summers, Capt. F. W., V-4
Suter, Dr. Geo., I-7
Svirbely, J. L., VIII-3:4
Swanson, Dr. C., VIII-3:3
Swanson, Maj. Melvin O., I-12, V-3
Swartout, Maj. C. W., VII-1
Sweeney, Maj. C. W., VIII-2
Swenson, E. E., IV-6
Swope, Herbert B., I-4:7
Szilard, Dr. Leo S., I-1, 4:4, 4:12, 4:13,

Taber, Rep. John, I-4:1
Tait, Capt. R., I-14
Talliaferro, Dr. W. H., I-7
Tallman, C. C., IV-1
Tammaro, A., Lt. Col., II-3, VIII-3:7
Tannenbaum, Dr. Albert, I-7
Tashek, R., VIII-2
Tate, Dr. John T., I-4:2
Taub, J. M., VIII-2
Taylor, E. H., VIII-3:7
Taylor, Fred, V-3

Taylor,G.I.,VIII-2
Taylor,Geo.W.,I-8
Taylor,Dean H.S.,I-4:2,II-2,III
Taylor,Lt.R.A.,Jr.,I-14,VIII-2
Taylor,Dr.T.T.,I-4:14
Teeple,Capt.D.S.,I-14
Teidje,H.F.,III
Telchow,Dr.,I-14
Teller,Dr.Edward,II-5,VIII-2
ten Bosch,M.,VIII-3:9
Tendam,D.J.,VIII-3:9
Tenney,T/Sgt.G.H.,VIII-2
Terrell,Dr.H.M.,I-7
Terry,Dr.L.L.,I-4:6
Tetenbaum,M.,VIII-3:7
Thacker,O.,Jr.,I-10
Thayer,Hanford,I-7
Thayer,H.E.,VII-1
Thayer,Dr.V.B.,III
Thiele,Dr.E.W.,III,VIII-3:7
Thomas,Dr.Chas.A.,I-4:2,4:3,4:4,
    4:7,VIII-2
Thomas,C.W.,VIII-2
Thomas,Sen.Elmer,I-4:1
Thomas,J.L.,VIII-3:7
Thomas,T.,VIII-3:7
Thomas,Capt.W.,VII-1
Thompson,Dr.A.F.,Jr.,I-4:5,4:7
Thompson,Lt.F.S.,I-14
Thompson,Maj.G.C.,I-14
Thompson,Capt.G.P.,VII-2
Thompson,H.E.,II-5
Thompson,Dr.J.G.,I-4:12,VII-1
Thompson,J.J.,II-2
Thompson,Dr.L.B.VIII-3:8
Thompson,LaRoy,VIII-2
Thompson,Dr.L.T.E.,VIII-3:2,3:9
Thompson,W.I.,II-3
Thompson,V.E.,VI
Thomsett,Capt.R.E.,VIII-1
Thorley,I.A.,VIII-3:4
Thornton,T/Sgt.G.,VIII-2
Thornton,Dr.R.L.,I-4:5
Thouren,Cmdr.E.H.,VIII-3:2
Thumser,R.C.,I-6
Tibbets,Col.P.W.,VIII-2,3:2,3:9
Tidemann,David,I-7

Titterton,E.W.,VIII-2,3:8
Titus,Lt.J.L.,I-14
Tolman,Dr.R.C.,I-1,4:2,4:4,4:5,4:7,
    4:13,4:14,VIII-2,3:4,3:8
Trask,C.H.,IV-5
Traver,Lt.R.W.,I-14
Travis,Maj.J.E.,I-5
Traynor,Maj.H.S.,III
Treon,Dr.J.F.,I-7
Tribby,Jas.,VIII-2
Truman,Pres.Harry S.,I-1,4:1,4:2,4:7,
    4:13,9,IV-1
Trump,Dr.,VIII-2
Trytten,M.H.,VIII-2
Tsuzuki,Dr.M.,I-4:6
Tuck,Dr.J.L.,VIII-2,3:8
Tucker,Ens.,VIII-2
Tucker,H.G.,I-12
Tucker,Tommy,I-12
Turner,C.D.,VIII-3:7
Turner,G.H.,VIII-3:7
Turner,Maj.J.H.,I-9
Tuve,Dr.M.A.,I-4:3,4:12,VIII-3:9
Twing,W.D.,II-4
Tybout,Lt.R.A.,I-4:6,7
Tyler,Col.G.R.,VIII-1,2
Tyler,W.T.,III

Uanna,Maj.W.L.,I-4:6,14
Ubelacker,A.A.,II-5
Ulam,S.,VIII-2
Ullman,R.,VIII-3:7
Ullrich,Lt.F.W.,USN,I-4:6
Underhill,R.M.,VIII-2,3:1
Urban,J.J.,IV-6,VIII-3:7
Urey,Prof.H.C.,I-4:5,4:7,4:12,4:13,
    4:14,II-2,III,VIII-2,3:3,3:7

Valente,Maj.F.A.,IV-2II,6
Valentine,Capt.Wm.,I-7
VanArtsdalen,E.R.,VIII-2
Vance,Lt.Col.John E.,I-4:7,14,VII-1
Van de Graaff,Dr.R.J.,VIII-2

VandenBulck,Lt.Col.Charles,I-5,6
VanDorn,R.M.,I-6
VanFleet,J.R.,VII-2
VanGemert,R.J.,VIII-2
VanHorn,Maj.E.L.,VII-1
VanHoy,Maj.J.W.,IV-6
VanKeuren,Rear Adm.,VI
VanVleck,Dr.J.H.,I-1,VIII-2
VanWie,N.H.,II-5
Varley,Capt.N.,I-4:6,II-1,VIII-3:4
Vaughan,F.B.,IV-6
Vaughan,Lt.Col.Harry,I-4:1
Vaughn,Henry,I-7
Vaughan,Lt.S.J.(3d),I-14
Velten,Capt.E.M.,VII-1
Vernon,Dr.H.C.,I-4:4
Vettel,Lt.C.T.,I-14
Vier,Dr.I.,VIII-2
Vinciguerra,Lt.J.B.,I-14
Vinson,Fred M.,I-8
Voghtlin,Dr.Carl,I-7
Voight,Adolf,I-4:11
Volkel,Dr.,I-14
Volkoff,Prof.G.M.,I-4:9
Vollmer,Lt.J.F.,VIII-1
Volpe,Joseph,I-4:7,14
Vona,J.A.,VIII-3:7
VonGrosse,Dr.A.,II-2
VonLaue,Prof.Max,I-14
vonNeumann,John,VIII-2,3:8
VonWeizsacker,Prof.C.F.,I-14

Wadkins,Robt.H.,I-12
Wahl,A.C.,IV-1,2I,VIII-2
Wake,Capt.D.L.,I-4:6
Waldman,B.,VIII-2
Waldmann,Dr.,I-14
Waldner,Lt.P.F.,I-14
Walker,M/Sgt.A.A.,I-4:6
Walker,R.L.,VIII-2
Wallace,Vice Pres.Henry A.,I-1,IV-1
Wallace,Dr.W.E.,VIII-3:3
Wallgren,Sen.Mon C.,I-4:1
Walsh,Lt.Harry R.,I-14
Walsh,Lt.Col.R.J.,Jr.,VII-1

Walter,Lt.H.P.,VII-1
Walters,F.M.,Jr.,VIII-2
Walther,Lt.M.M.,USN,VIII-3:2
Wantman,Dr.Maury,I-7
Ward,W.H.,IV-1
Wardenburg,Dr.F.A.C.,I-14,III
Ware,Capt.J.T.,I-9,VIII-3:2
Warf,Jas.C.,I-4:11
Warner,J.C.,VIII-3:4
Warner,E.S.,Jr.,VIII-2,3:6,3:8
Werner,Lt.W.L.,I-14
Warren,L.G.,V-6
Warren,Capt.Shields,USN,I-4:6
Warren,Col.Stafford L.,I-4:3,4:6,7,
    VIII-2,3:4,3:8
Waterman,T.E.,I-6
Watson,Dr.C.J.,I-4:3,7
Watson,Prof.E.C.,VIII-3:2
Watson,Dr.W.W.,I-4:4,4:9
Webb,Dr.G.M.,III
Webb,Capt.Martin,I-14
Webb,W.L.,VIII-3:7
Weber,E.C.,IV-2II
Webster,Col.,I-4:6
Wedemeyer,Lt.Gen.A.C.,VIII-3:8
Weiboldt,Dr.J.,VIII-3:8
Weidenbaum,B.,VIII-2
Weil,Dr.Geo.L.,I-4:9
Weimer,H.R.,VIII-3:4
Weimer,Dr.Karl,I-14
Weinberg,Dr.Albert,I-4:4
Weinberg,Dr.A.M.,I-4:7
Weingarten,Lt.J.L.,I-14
Weisner,J.B.,VIII-2
Weiss,Dr.H.,VIII-3:8
Weiss,T/3 Mildred,VIII-2
Weisskopf,V.F.,VIII-2
Weissman,S.I.,VIII-2
Welch,Capt.Wm.,VIII-1
Wells,Lt.Algie A.,I-14
Wells,P.A.,I-9
Wells,Capt.W.G.,III
Welsh,Col.A.B.,I-7
Welton,T.,VIII-2
Wendt,Lt.C.F.,I-14
Wensel,Dr.H.T.,I-4:12,4:13,13
Wentworth,R.A.,IV-2II

~~CONFIDENTIAL/RD~~   SECRET

Werner, H. E., IV-5
Westhaver, Dr. J. W., I-4:14
Wetherhold, Dr. J. M., I-7
Weygand, Prof. Fritz, I-14
Wheeler, Dr. J. A., I-4:2, 4:4, IV-1, 6
Whipple, Capt. Harry O., I-4:6, VIII-2
Whitaker, Capt. Albert E., Jr., I-14
Whitaker, M. D., IV-2II
White, Capt. Chas. V., I-14
White, Maj. E. A., I-9, VIII-1, 2
White, Lt. H. R., I-14
White, J. G., V-6
White, W. A., VII-2
White, Sen. Wallace H., Jr., I-4:1
Whitehurst, B. W., V-3
Whittaker, Dr. M. D., I-4:3, 4:4
Whittelsey, C. C., II-4
Wichers, E., VIII-2
Wick, Dr., I-14
Wiedehecker, K. P., V-5
Wien, W., II-2
Wiener, W. W., I-5
Wier, Maj. R. J., I-9
Wiesner, Dr. J., VIII-3:8
Wietig, Dr., I-14
Wigner, Dr. E. P., I-1, 4:2, 4:4, 4:7,
    4:9, 4:13, IV-2II
Wilear, V-6
Wilhelm, Dr. H. A., I-4:11
Wilhoyt, Lt. Col. E. E., VIII-2, 3:6, 3:8
Wilker, A. V., II-5
Wilkinson, P. G., VIII-3:3
Willard, Dr. J. E., I-4:4
Williams, Maj. Gen., VIII-3:9
Williams, Dr. Clarke, I-4:7
Williams, Lt. Col. Donald G., I-14
Williams, Lt. Fred T., USN, I-13
Williams, Dr. John, VIII-3:8
Williams, J. H., VIII-2
Williams, R., II-3, IV-1, 6
Williams, Col. W. J., I-4:7, II-1, 5
Willingham, W. A., I-5
Wilson, C. E., IV-1
Wilson, E. B., VIII-2
Wilson, F. W., IV-6
Wilson, Fld. Mar. Sir Henry Maitland,
    I-4:13

Wilson, J. D., IV-2II
Wilson, Col. Rosco O., I-4:6, VIII-2, 3:9
Wilson, Dr. R. R., I-4:14, VIII-2
Wilson, V. E., VIII-3:2
Wilt, D. L., VIII-2, 3:1
Winkelman, D. W., II-4
Winne, Harry A., I-4:7
Winstead, Capt. Chas. B., I-14
Winters, C. E., V-2, VII-1
Wirth, Dr. H. E., VIII-3:3
Wirtz, Dr. Karl, I-14
Wisansky, W., VIII-3:7
Wischmeyer, R. R., II-4
Wise, Capt. Harry, VIII-1
Wiesner, R. R., V-3
Woernley, D. L., VIII-3:4
Wolf, Capt. B. S., I-4:6, 7, VII-1, VIII-3:4
Wolff, Lt. Col. Charles, I-9
Wollan, Dr. E. W., I-7
Womeldorff, F. M., I-9
Wood, M. F., IV-2II, 5
Woods, W. K., IV-6
Workman, E. J., VIII-2
Worth, Dr. J. E., I-7
Worthington, Hood, I-4:4, IV-6
Wright, Dr. G. E., III
Wright, C. W., IV-4
Wright, O. L., VIII-2
Wroe, Lt. M., VIII-1
Wulff, Lt. A. V. R., Jr., I-14

York, Maj. D. L., I-5
Young, I-4:9
Young, Dwight, VIII-2
Young, Dr. Gale, I-4:4, 4:7
Young, E. R., V-5
Youngs, H. S., I-4:3
Youngs, Maj. W. C., Jr., I-4:6, 12

Zacharias, Prof. J. R., I-4:2, VIII-2
Zeitlin, F., VII-1
Zimmerli, T/4 F., VIII-2
Zindle, Lt. H. J., I-14

~~CONFIDENTIAL/RD~~

SECRET

RESTRICTED DATA
ATOMIC EN████ ACT 19██

Zinn, Dr. W. H., I-4:2, 4:13, VIII-2
Zirkle, R. E., I-4:3, 7
Zmachinsky, W. C., VIII-3:7
Zumwalt, Capt. L. R., V-6
Zussman, Dr. Bernard, I-7

ADDENDA

Barr, Gen. D., I-14 S2
Bull, Maj. Gen. H. R., I-14 S2

Castles, Col. J. W., I-14 S2
Conrad, Col. G. B., I-14 S2

Devers, Gen. J. L., I-14 S2

Eisenhower, Gen. Dwight D., I-14 S2
Eyster, Col. G. S., I-14 S2

Fell, Col. E. T., I-14 S2

Gruver, Col. E. S., I-14 S2

Hawley, Gen. P. R., I-14 S2

Jervey, Col. W. W., I-14 S2

O'Leary, Mrs. Jean, I-1, 3.15
Olsson, Miss Virginia, I-1, 3.14

Redlinger, Capt. M. J., I-14 S2
Rugg, Lt. George A., I-14 S3

Sibert, Gen. E. L., I-14 S2
Smith, Gen. W. B., I-14 S2

Timothy, Col. P. H., I-14 S2

Whiteley, Gen. J. F. M., I-14 S2

C(2) <u>INDEX OF NAMES OF</u>

<u>AGENCIES, INDUSTRIAL ORGANIZATIONS, UNIVERSITIES, ETC.</u>

<u>Notes:</u>

1.   For explanation of references, see Notes 1, 2 and 4, preceding previous Index, C(1), of Names of Persons.

2.   The names of Manhattan District installations and offices, and of many Army and Navy organizations, and of other Government agencies whose contacts with the Manhattan District were for the most part of a routine nature, have been omitted from this Index.

3.   Some names, not included in the main index, have been inserted as "Addenda", on p. C(2) 14.

APPENDIX D

MAP, MANHATTAN DISTRICT INSTALLATIONS AND OFFICES

MANHATTAN DISTRICT INSTALLATIONS AND OFFICES

www.ingramcontent.com/pod-product-compliance
Lightning Source LLC
Chambersburg PA
CBHW050619110426
42813CB00010B/2612

9 781608 881765